LOGIC & NATURAL LANGUAGE

Logic & Natural Language

On Plural Reference and Its Semantic and Logical Significance

HANOCH BEN-YAMI
Tel Aviv University

Routledge
Taylor & Francis Group

LONDON AND NEW YORK

First published 2004 by Ashgate Publishing Limited

2 Park Square, Milton Park, Abingdon, Oxfordshire OX14 4RN
52 Vanderbilt Avenue, New York, NY 10017

Routledge is an imprint of the Taylor & Francis Group, an informa business

First issued in paperback 2019

British Library Cataloguing in Publication Data
Ben-Yami, Hanoch
    Logic & natural language : on plural reference and its
    semantic and logical significance. - (Ashgate new critical
    thinking in philosophy)
    1.Language and logic 2.Predicate calculus 3.Grammar,
    Comparative and general - Quantifiers 4.Semantics
    (Philosophy)
    I.Title
    121.6'8

Library of Congress Cataloging-in-Publication Data
Ben-Yami, Hanoch, 1962-
    Logic & natural language ; on plural reference and its semantic and logical
significance / Hanoch Ben-Yami.
        p. cm.
    Includes bibliographical references and index.
    ISBN 0-7546-3743-3 (alk. paper)
    1. Grammar, Comparative and general--Number. 2. Grammar, Comparative and
general--Quantifiers. 3. Reference (Linguistics) 4. Language and logic. I. Title:
Logic and natural language. II. Title.

P240.8.B46 2004
302.23'01--dc22

                                                              2003063718

ISBN 13: 978-0-7546-3743-1 (hbk)
ISBN 13: 978-1-138-27380-1 (pbk)

Contents

Preface   vii

1. Introduction   1

PART I: PLURAL REFERRING EXPRESSIONS

2 Plural Referring Expressions in Natural Language   7
   2.1 The Common View on Reference   7
   2.2 Plural Reference   8
   2.3 The Implausibility of Reductive Analyses of Plural
        Referring Expressions   16

3 Common Nouns as Plural Referring Expressions   28
   3.1 The Functioning of Common Nouns   28
   3.2 On an Alleged Ambiguity of the Copula   31
   33   Attributive and Predicative Adjectives   32
   3.4 Natural Kind Terms   35
   3.5 Empty Names   37

4 The Sources of the Analysis of Referring Nouns as Predicates   41
   4.1 Frege   41
   4.2 Russell and Bradley   44

5 Reference   47

PART II: QUANTIFICATION

6 Quantification: Natural Language versus the Predicate Calculus   59
   6.1 The Nature of Quantification, and the Differences between
        Its Implementations   59
   6.2 Definite and Indefinite Noun Phrases   61
   6.3 Geach and Strawson on Plural Reference and Quantification   63
   6.4 Binary and Restricted Quantification, and Comparative Quantifiers 66
   6.5 Is 'Existence' a Quantifier?   73

7 Multiple Quantification   78
   7.1 On Ambiguity and Formalization   78
   7.2 Iterative Reading of Multiply Quantified Sentences   80
   7.3 Additional Readings of Quantified Sentences   84
   7.4 On the Passive, Converse Relation-Names, and the Copula   90

8   Pronouns, Variables, and Bound Anaphors   95
     8.1   Pronouns and other Definitive Noun Phrases as Alleged Variables   95
     8.2   Variables versus Bound Anaphors   96
     8.3   Rules for the Choice of Anaphors   98
     8.4   Conditional Donkey Anaphora   100
     8.5   Predicate Connectives, and Bound Anaphora across
           Sentential Connectives   107
     8.6   The Relation between the Truth-Value of a Quantified Sentence
           and those of Its Instances   109

PART III: A DEDUCTIVE SYSTEM FOR NATURAL LANGUAGE

9   Derivation Rules and Consistency   115
     9.1   Some General Considerations   115
     9.2   Basic Characteristics of the System   117
     9.3   Transposition   119
     9.4   Universal Elimination   120
     9.5   Universal Introduction   121
     9.6   Particular Introduction   122
     9.7   Particular Elimination   123
     9.8   Referential Import   124

10   Applications I: Aristotelian Logic   128
     10.1 The Square of Opposition   128
     10.2 Immediate Inferences   130
     10.3 Syllogisms   131

11   Application II: Beyond Aristotelian Logic   133
     11.1 Generalization of Transposition   133
     11.2 Multiply Quantified Sentences   134
     11.3 Predicate- and Sentence-Connectives   136
     11.4 The Logic of Relations   140
     11.5 Identity   142

12   Conclusions   148

Bibliography   153
Index   158

Preface

As far back as I can remember, I have always had doubts concerning the adequacy, in some sense or other, of the predicate calculus. The central idea of this book, however – that the predicate calculus lacks plural referring expressions, and that consequently it incorrectly construes common nouns as always predicative – came to me sometime in late 1995 or early 1996. Since then I have intermittently developed this idea, and others to which it gave rise, initially thinking that I would be able to publish my conclusions in a long paper. My invention in the summer of 2000 of the deductive system developed in Part III below, made me realize that my work would have to be published as a monograph. Since then I have been working more or less continuously on this book.

I presented parts of my work in departmental colloquia in Israel during the academic year of 1996-7, and in the Second European Congress of Analytic Philosophy in Leeds, in 1996. Although I received a few helpful comments on these occasions, it was clear that both the time allocated for presentation on such events and the state of my work made it unsuited to them.

I gave three seminars on my work at Tel Aviv University, which were very helpful. I am much indebted to my students in these seminars, who are however too numerous to be named here.

Several friends and colleagues have read this work or parts of it at different stages of its development, and have made comments that significantly contributed to its improvement. Maria Alvarez read one chapter, as did Roger Teichmann. Stephen Blamey read Part III, which contains my deductive system; his comments and our discussions were very helpful, especially for my discussion of identity. Eli Dresner and Bede Rundle read my entire manuscript. My student Ran Lanzet has also read the entire manuscript, some parts more than once, and our discussions and correspondence were invaluable in many respects. I am similarly indebted to Peter Hacker, whose influence on my philosophizing also extends beyond the limits of this work.

My greatest debt is to John Hyman. John too read my entire manuscript, parts of it several times, and his numerous comments, on every possible aspect, were always extremely helpful. But John's friendship contributed to this work in other ways as well. Among other things, he made possible my sojourns in Oxford; these contributed in several ways to my philosophical development, in which our endless exchanges played a major role. Without John's help and encouragement, this work would have been of much inferior quality, if it had existed at all.

Because of its very nature, the literature relevant to this work is enormous. I am therefore convinced that I have failed to notice some works which are pertinent to some of the topics discussed in this book. I would be grateful to anyone who draws my attention to such works. However, since my basic claims concern subjects that are at the foundation of any logical or semantic theory, it is unlikely that I have overlooked any influential work on these matters.

The ideas presented in this book can be developed in additional directions. Their application to modal logic, for instance, and the corresponding expansion of the deductive system of Part III, are rather straightforward. Nevertheless, I decided to avoid any such further investigations in this book. I wanted it to be a focused argument concerning one fundamental issue. The point of any further developments depends on the acceptability of my basic claims. Ramifications of these claims may be pursued at a later stage.

I am not a native English speaker, and writing in this language has been a constant struggle, in which I am afraid I wasn't always the winner. I have also been continuously occupied with other projects beside this one. For these reasons, and others that concern my capabilities, I believe this book could be better written. Yet perhaps it succeeds in making its basic claims compelling. The fact that it criticizes accepted positions, at the core of contemporary philosophy of language and logic, might cause some to dislike it; but, I hope, only some.

H.B.
Tel Aviv, January 2004.

Chapter 1

Introduction

Frege's invention of the predicate calculus, first published in his *Begriffsschrift* of 1879, has been the most influential event in the history of modern logic. The predicate calculus made formal logic an object of study for many logicians and mathematicians, and consequently fundamental logical concepts were clarified and some notable logical theorems were proved. Moreover, the calculus both constituted a language which many found of philosophical significance, and was used for the analysis of natural language. Partly as a result, the philosophy of language acquired an unprecedented eminent position in philosophy. For some time, many philosophers thought that it is, or should be, what philosophy consists in; and although this position is not prevalent any longer, the importance of logic and the philosophy of language both as domains of philosophical investigations and for other philosophical studies remains undisputed.

The predicate calculus' place in logic is so central that many philosophers, when they think of logic, think in fact of the calculus. Since Frege's time the calculus has undergone several modifications, but all versions studied and used today are very close descendants of his *Begriffsschrift*. Any introductory course on logic devotes much of its time, and frequently most of it, to the study of the calculus. For many, to speak about logic is to speak about the calculus.

Moreover, logicians, philosophers and linguists analyze natural language by means of the calculus. When one writes on the logical form of an ordinary sentence, one means by it the form of the sentence's translation into some version of the calculus. When one analyzes the semantics of natural language, one does it with the apparatus of sentential functions, quantifiers, variables, domain of discourse, scope, etc., all borrowed from the calculus and exemplified in its formulas.

It is, indeed, usually admitted that Frege's calculus is insufficient for the adequate analysis of the semantics of natural language; but, it is thought, the calculus needs only to be enriched in various respects in order to become adequate for that purpose. The following paragraph from Wiggins (1997, p. 5) is representative of the current attitude:

> Given the universality and generality of the insights that originate with Frege, what we now have to envisage is the final extension of *Begriffsschrift*, namely the extension which, for purposes rather different from Frege's, will even furnish it with the counterpart of such ordinary sentences as "the sun is behind cloud" (say). In the long run, the extended *Begriffsschrift* might itself be modified further, to approximate more and more closely to the state of some natural language.

I think that the calculus' position in contemporary logic and philosophy of language is based on mistaken assumptions. In order for the calculus to be used as a tool for the study of natural language, its semantic categories should parallel those of the latter and be implemented in the same way. But I hope to show in this book that neither is the case. Most importantly, I shall argue that the way reference is incorporated in the calculus is fundamentally different from the way it is incorporated in natural language, and that as a result predication and quantification in the two systems are profoundly dissimilar. Consequently, reference, predication and quantification in natural language cannot be understood if one attempts to explain them by means of the calculus. Thus, the logic and semantics of sentences of natural language cannot be captured by the calculus. One distorts the semantics and logic of natural language when one studies them by means of the calculus.

By contrast, the calculus, as a language with a semantics and logic of its own, is a legitimate object of study for Logic. And I do not think that the calculus involves any incoherence. Of course, the question then arises: if the calculus cannot contribute to our understanding of natural language, why should its study be of interest? Shouldn't Logic investigate the languages actually used in the various fields of knowledge? We shall return to this question in the conclusion.

A main purpose of this book is, therefore, to demonstrate several significant semantic distinctions between the predicate calculus and natural language, distinctions that make the former inadequate for the study of the semantics and logic of the latter.

In order to accomplish that, I pursue various logical and semantic investigations of natural language. Some of these investigations may posses, however, independent interest as well. For instance, the system of natural deduction for natural language, developed in Part III, may be found interesting, independently of its contribution to the criticism of the value of the predicate calculus to the study of natural language. Accordingly, these investigations of semantic and logical properties of natural language can be considered independently of their contribution to the critical purpose of this work.

The book is divided as follows. In Part I I discuss plural reference. I explain what plural reference is, and I show that natural language, in contrast to the predicate calculus, uses plural referring expressions. Most significantly, I argue that common nouns, in many of their uses, are such expressions. I consider Frege's and subsequent arguments to the contrary, and show them to be unsound; this leads to a discussion of the nature of reference.

In Part II I discuss the nature of quantification. I contrast quantification in the predicate calculus with quantification in natural language, and I show that the absence of plural referring expressions from his calculus made Frege introduce quantification into it in a way significantly dissimilar to the way quantification functions in natural language. I continue to show how my analysis applies to multiply quantified sentences and how it avoids difficulties that confront the predicate calculus, even in its versions that employ generalized quantifiers. I explain the logical necessity of devices like the passive, converse relations and the copula for natural language – all logically redundant from the predicate calculus' point of view. I also discuss anaphora, in order to show in what way bound anaphors are different from bound variables, and how bound anaphora functions across sentential connectives.

In Part III I develop, on the basis of the semantic foundations laid in the previous parts, a deductive system for natural language. I first introduce my derivation rules and prove the consistency of my system. I then prove the valid inferences of Aristotelian logic – the Square of Opposition, immediate inferences and syllogisms. I proceed to prove some logical relations between multiply quantified sentences and some properties of relations. I conclude this part by incorporating identity into my system.

A few preliminary remarks are necessary before we proceed. Firstly, although the semantics usually applied today to the predicate calculus is model-theoretic semantics, my discussion in this work is not committed to this interpretation of the calculus. This is for several related reasons. Firstly, Frege, as well as Russell and other developers of the calculus, did their work before model-theoretic semantics was invented by Tarski. It might therefore be unwarranted to commit their work to a semantics with which they were not familiar. Moreover, their conception of meaning, in so far as they had any general theory of meaning, does not always agree with model-theoretic semantics. For instance, Frege's distinction between *Sinn* and *Bedeutung* is not incorporated into that extensional semantics. So a criticism of the predicate calculus as interpreted by that semantics might be based on features to which the calculus is not necessarily committed.

In fact, if we consider how sentences of natural language are translated into the calculus, we see that the calculus is committed by that translation to something quite minimal. Its predicate letters should be interpreted as predicates (hence their name), its singular constants as singular referring expressions, its sentence connectives as sentence connectives, and its quantified constructions as parallel to those of natural language. We can accordingly regard predication in the calculus, for instance, as the same as that in natural language, without committing ourselves to any theory of predication. Any further theory of predication should be equally applicable to both languages. Model-theoretic semantics is just one such theory, problematic in several respects (e.g., in being extensional). We should not limit the calculus to the way this theory construes meaning.

Furthermore, the way meaning is sometimes interpreted by means of model theory in contemporary linguistics is definitely unacceptable. For instance, a proper name – say 'John' – is claimed to denote the set of all subsets of the domain of discourse that contain John as a member. Assuming that we have any understanding of what we mean when we say, e.g., that John is asleep, this claim is surely mistaken.

I shall not, therefore, rely on model-theoretic semantics in this work. In particular, my discussion in Part II of the achievements of modern formal linguistics will ignore what I take to be the inessential contribution of this semantics. I shall discuss model-theoretic semantics only once (note 2, page 31), where it might seem to offer a reply to one of my criticisms.

My second preliminary remark concerns my use of technical terms. Although I occasionally use terms which are found in the literature, I do not always observe their accepted meaning. This is for two reasons. Firstly, some are used in different ways by different authors – in such cases, an accepted meaning does not actually exist. Secondly, their explanations or use often presuppose semantic theories to which I wish not to be committed. I therefore explain most of the technical terms I introduce, even when my use of such a term agrees with some use found in the literature. My claims should always be judged on the basis of these explanations.

A related point is that I attempted to make this work accessible to students whose relevant background consists of standard introductory courses in philosophical logic and philosophy of language. I therefore introduce, although concisely, some material with which many readers will be familiar.

My third and last preliminary remark concerns my attempt to contrast the semantics of the predicate calculus with that of natural language. There are *many* natural languages, some quite different from each other in many respects. It therefore seems that one should discuss the semantics of this or that natural language, not the semantics of natural language in general. And this may cast doubt on the coherence of my project.

However, the semantic properties I am about to ascribe to natural language are such that should be expected if language is to be an efficient tool for describing things and events of importance for us – and all natural languages are very efficient at that. To give an example, unrelated to my discussion below, of such a semantic property: although the grammar of tenses differs widely between some languages (as it does between English and Hebrew, say), we should expect all natural languages to have the means to distinguish between an event being past, present or future. The semantic properties I discuss in this paper are true of all languages I have checked, which include such quite dissimilar languages as Indo-European languages, Semitic languages and Chinese. For this reason I believe I am justified in contrasting the predicate calculus with natural language generally, without maintaining that all natural languages are semantically equivalent in each and every detail.

All the same, it is still possible that some of the semantic observations I make below do not apply to some natural languages. This should be empirically determined, and I have checked but a negligible number of the world's thousands of languages (although I have sampled a non-negligible number of language families). As I said, constraints that have to do with the descriptive power of language make me inclined to reject this possibility. But even if I am mistaken, the fact that my claims are true of a wide variety of natural languages is sufficient for a criticism of the value of the predicate calculus to the study of the semantics and logic of natural language.

PART I

PLURAL REFERRING EXPRESSIONS

Chapter 2

Plural Referring Expressions
in Natural Language

2.1   The Common View on Reference

The predicate calculus distinguishes between singular terms, which are said to denote, designate, "stand for" or refer to particulars; and predicates, which can be said to attribute properties and relations to particulars denoted by singular terms. The form of the basic sentence of the calculus is '$P(a_1, \ldots a_n)$', where each of 'a_1' to 'a_n' designates a particular, and 'P' is used to say that an n-place relation holds between these particulars, or, in case $n=1$, that the single particular referred to has a certain property.

Philosophers and logicians who study the semantics of natural language apply this distinction to natural language as well. Firstly, the referring expressions of natural language are taken to be singular terms. This almost universal position is usually presupposed without being made explicit in logical discussions, but some philosophers do state it. Quine, for instance, writes that '[o]ne thinks of reference, first and foremost, as relating names and other singular terms to their objects.'[1] Gareth Evans writes about 'singular terms or referring expressions' and notes that 'these two phrases will be used interchangeably throughout' (1982, p. 1). He then elaborates (ibid., p. 2):

> In coupling a referring expression with a predicate, say 'smokes', a speaker intends to be taken to be making a remark about just one particular thing – a remark that is to be determined as true or false according to whether some one indicated individual smokes. So it is said that the role of a referring expression is that of indicating to the audience which object it is which is thus relevant to the truth-value of the remark.

And similarly, Stephen Neale writes (1990, p. 15; cf. his 1995, p. 765):

> With respect to natural language, I shall use 'genuine referring expression' (or 'genuine singular term') to cover ordinary proper names, demonstratives, and (some occurrences) of pronouns.

---

[1] At some stage in the process of revising this work I mistakenly deleted and consequently lost the reference of this quotation. I nevertheless use it here, since it succinctly expresses the position I am criticizing, and since it is supported, as representative of Quine's view, by the additional quotations from him below.

Logic & Natural Language

In consequence, philosophers maintain that predication, as Quine writes, 'joins a general term and *a singular term*' (1960, p. 96; italics added). We refer to a particular and say something about it.

Secondly, common nouns, even when in the grammatical subject position, are taken to be logical predicates, like adjectives and verbs. In a purported translation of sentences of natural language into the predicate calculus, all these parts of speech are translated by predicates.

Frege introduced this assimilation into modern logic already in his *Begriffsschrift*, by treating common nouns in the subject position as logical predicates. He translates the subject '*S*' in the four Aristotelian subject-predicate sentences, 'All/some *S* are/aren't *P*' by a predicate (1879, § 12). Similarly, he takes the common noun 'house' to be a predicate, and translates the sentence 'There is at least one house' into his calculus as 'Not(all *x*)Not(House *x*)', where 'House' is a predicate (1879, § 12, note). And predicate letters, Frege says, attribute properties to particulars (ibid.). The assimilation of common nouns to predicates reappears in all his later writing (see especially his 'Über Begriff und Gegenstand', pp. 197-8). Frege's analysis was almost unanimously accepted by later logicians.

Quine writes on the topic as follows (1960, p. 96; cf. his 1981, pp. 164-5):

> It happens that the separation of roles into those that call for the substantival form, those that call for the adjectival, and those that call for the verbal has little bearing on questions of reference. Our study can consequently be simplified by viewing substantive, adjective, and verb merely as variant forms given to a general term. Thus we may best picture predication in the neutral logical schematism '*Fa*', understood as representing not only '*a* is an *F*' (where '*F*' represents a substantive) but also '*a* is *F*' (where '*F*' represents an adjective) and '*a Fs*' (where '*F*' represents an intransitive verb). Predication is illustrated indifferently by 'Mama is a woman', 'Mama is big', and 'Mama sings'. The general term is what is predicated, or occupies what grammarians call predicative position; and it can as well have the form of an adjective or a verb as that of a substantive.

Linguists who study the semantics of natural language were influenced by formal logic in their analyses. Consequently, they too interpret common nouns and adjectives similarly, taking both to attribute properties to particulars (Keenan, 1996, p. 42).

Philosophers and logicians ascribe the referential function to singular terms, and take common nouns to be predicates, like adjectives and verbs.

2.2 Plural Reference

I think this prevalent view is mistaken. *Natural language, in contrast to the predicate calculus, has plural referring expressions as well.* And while the predicate calculus conflates the common nouns of natural language with its adjectives and verbs, treating all as predicates, *common nouns are often used not as predicates, but to refer to particulars.*

I shall try to establish these two claims in this part of my book. Common nouns as referring expressions will be the subject of the following two chapters; in the rest of this chapter I shall try to explain the nature of plural reference, and to establish the existence of irreducible plural referring expressions in natural language.

The nature of *reference* is complex (I shall discuss it to some extent in Chapter 5). By contrast, what is involved in *plural* reference, vis-à-vis *singular* reference, is, I think, straightforward. Whatever reference is, we can refer to a single person, say by means of a word. 'Who is living in this house?' – 'John': here we referred with a word, 'John', to a man, John. Now do that several times in a row, each time referring to a different person, and you have plural reference. For instance: 'Who is living in that house?' – 'John and Mary'. Or 'John and Mary have arrived' in contrast to 'John has arrived'. Whatever is achieved in referring to a single person or thing can be achieved with respect to several persons or things, and we then have plural reference.

When I talk about plural reference I mean referring to more than a single person or thing. I *don't* mean referring to a set with many members, to a complex individual, or to any other variation on these ideas. I mean achieving with relation to more than a single thing what is achieved by reference to a single thing.

In my examples, the plural referring expressions were conjoined into a plural noun phrase by the word 'and'. But it seems that this word's function is not to contribute to the manner of reference, but to indicate how the predicate – 'living in this house' and 'has arrived' in my examples – should apply to the persons or things referred to. If one replied to the question 'Who is living in this house?' by saying 'John *or* Mary' instead of 'John *and* Mary', still the same two persons would be mentioned or referred to, but this time the reply would be correct if at least one, and not necessarily both, were living in that house.

I used proper names in my examples of plural referring expressions, but other singular referring expressions are often used in these ways too. Pronouns such as 'I' or 'she', demonstratives like 'this man' or 'that woman', and definite descriptions like 'the man over there' or 'my sister' should or could be similarly used in appropriate circumstances. And I conjoined only two referring expressions for the sake of brevity alone. Longer conjunctions are often used, and the limitations in practice on the length of such lists seem not to arise out of any syntactic or semantic principle.

In each of my examples of plural reference I conjoined two words, each referring to a single person, to achieve plural reference. But once it is admitted that this kind of reference can be achieved by several singular referring expressions in a row, there is no reason why it could not be achieved by a single *plural* one. 'Who is living in this house?' – 'The Browns', 'They do', 'These people', 'The people who have just walked past us': here reference to several people, who might be John and Mary of our previous examples, is achieved by a single expression or word. (Although a plural referring expression is in most cases a plural noun phrase, these semantic and grammatical categories do not coincide: 'trousers', a plural noun, is often used as a singular referring expression.)

It thus seems *prima facie* plausible to maintain that plural pronouns ('we', 'you', 'they'), plural demonstratives ('these', 'these books', 'those pictures' etc.) and plural definite descriptions ('the books you gave me', 'my books', etc.) are often used to refer to more than a single thing. (All these expressions may have other, non-referential uses, as well.) Additionally, expressions like 'the Browns', which are grammatically plural definite descriptions, seem to function occasionally somewhat like a plural proper name.

Moreover, the use of singular referring expressions is paralleled very closely by the use of plural expressions – I shall describe these parallels in some detail in Chapter 5, where I discuss reference generally. This parallelism supplies us with a very good reason to consider these uses of plural expressions referential.

Of course, this is not the way such expressions are usually construed by contemporary philosophers and linguists. Since in the predicate calculus the referential function is limited to singular referring expressions, most philosophers and linguists, as has been demonstrated in the previous section, think of reference as essentially singular. Accordingly, one usually tries to reduce apparent plural referring expressions like 'we', 'these men' or 'my sisters' to the semantic categories recognized by the calculus. I shall examine the justification and success of these attempts in the next section. Yet, once it is admitted that plural reference can be achieved by conjunctions of singular referring expressions, there is no reason why one cannot *stipulate* that a given expression, say 'the Browns', can also be used as a plural referring expression, designating John and Mary Brown. And if such a stipulation is possible, and moreover natural and useful, then it seems *prima facie* implausible that this is not a way in which this expression, and probably expressions of other kinds as well, are actually used in natural language.

As I have just noted, most philosophers and linguists who have discussed plural referring expressions usually try to reduce them to other kinds of expression. There are, however, some who have not considered such a reduction necessary, the most notable of which are perhaps Peter Strawson and Peter Geach. My analysis of quantification being similar to Strawson's and Geach's in some respects, I shall postpone the comparison of our views to Part II of my book (§ 6.3), in which quantification is discussed. I shall also discuss there (p. 85), for similar reasons, Boolos's semantics of second-order logic, which involves plural reference. Here I shall mention a few other philosophers who have discussed plural reference in their works. This will serve both to elucidate the idea of plural reference and to emphasize some distinctive features of my claims.

I start with Max Black, who may have been influenced by Geach, with whom he was well acquainted. Black noted very clearly the existence and nature of plural referring expressions in ordinary language (1971, pp. 628-9):

> The most obvious ways of referring to a single thing are by using a name or a definite description: 'Aristotle' or 'the president of the United States'. Equally familiar, although strangely overlooked by logicians and philosophers, are devices for referring to several things *together*: 'Berkeley and Hume' or 'the brothers of Napoleon'. Here, lists of names (usually, but not necessarily, coupled by occurrences of 'and') and what might be called "plural descriptions" (phrases of the form 'the-so-and-so's' in certain uses) play something

like the same role that names and singular descriptions do. Just as 'Nixon' identifies *one* man for attention in the context of some statement, the list 'Johnson and Kennedy' identifies two men at once, in a context in which something is considered that involves both of them at once. And just as 'the President of the United States' succeeds in identifying one man by description, so the phrase 'the American presidents since Lincoln' succeeds in identifying several, in a way that allows something to be said that involves all of them at once.

Black proceeds to demonstrate how natural the concept of plural reference is:

> The notion of "plural" or simultaneous reference to several things at once is really not at all mysterious. Just as I can point to a single thing, I can point to two things at once – using two hands, if necessary; pointing to two things at once need be no more perplexing than touching two things at once.

My conception of plural reference resembles Black's. I do not think, though, that plural reference must be compared to the kind of simultaneous pointing that Black mentions. When we say 'Berkeley and Hume', the names are uttered consecutively, and not simultaneously – we can similarly point to several people by pointing to them one after another. We can also point with a single gesture to several people standing close to each other.

Black adds, however, only very little to what I have just quoted; the radical consequences of plural reference to logic and semantics that I discuss below were missed by him. Moreover, Black mentions only plural definite descriptions and conjunctions of names as examples of plural referring expressions. The use of common nouns in quantified noun phrases to refer to many things at once, which I discuss in the next chapter, seems to have eluded him.

Peter Simons has also made use of plural reference in his discussion of the nature of number and of set theory (Simons, 1982, to which are all page references below). On his work on sets he writes that it is 'in a large part a development of the line of thought opened up in particular by Black [he refers to Black's 1971 paper], who first formulated with clarity the view that sets are to plural terms as individuals are to singular terms' (p. 200). By plural terms Simons means 'the sort of expression which can be used to refer to more than one thing at once' (p. 165). He counts as plural terms plural definite descriptions, plural demonstratives, plural personal pronouns, name lists ('Tom, Dick and Harry'), mixed term lists ('Jason and the Argonauts'), and – 'if there are any' – plural proper names (ibid.). Simons calls the several individuals designated by a plural term a *manifold*, and he emphasizes that 'there is no difference between the manifold, and the several individuals, despite the fact that we can talk about *a* manifold' (p. 166). Simons specifies parallels between singular and plural terms (ibid.):

> Just as one and the same expression which is a singular term may on different occasions of its use denote different individuals, so one plural term may also on different occasions designate different manifolds. Just as 'the President of the United States' denotes different

men at different times, so 'Farmer Brown's prize herd of Friesians' may on different occasions designate different manifolds of beasts. Similarly, just as two terms with different meanings may yet have the same referent when singular, so two plural terms with different meanings may yet have the same referents.

I am in agreement with Simons about these features of plural reference. As I noted above, among plural referring expressions I consider also lists of singular referring expressions connected by a disjunction, and not only by a conjunction – the connective indicates the way the predicate applies to the particulars referred to, and does not affect the manner of reference. But this is a minor point. The essential difference between me and Simons is that while I consider common nouns, in some of their uses, plural referring expressions, Simons explicitly rejects this possibility. On 'general terms, such as "man", "hooded crow", "horse with a wooden leg" etc.', which he calls *common noun phrases* (CNPs), he writes (p. 206): 'I believe Frege was right in considering such general words and phrases ... as being inherently predicative rather than referential...'. And further on he adds: 'A plural term like "the people in this room" is to be sharply distinguished from the (plural) CNP "people in this room". ... Semantically, CNPs do not of themselves make definite reference to things' (p. 207). (Simons indeed writes that he does not 'consider CNPs to be *simply* predicates' (pp. 206-7), but this is mainly for syntactic reasons, as is clear from his note 29 to page 209.)

My rejection of Frege's analysis of common noun as predicative and their analysis as plural referring expressions will lead me below to a radically different analysis of quantification, as well as of many other features of natural language. By contrast, Simons' acceptance of Frege's analysis brings him to adopt Frege's predicate calculus. The formal language Simons develops (pp. 215-36) is not a *departure* from Frege's calculus but an *elaboration* of it – mainly by the introduction of plural terms and plural variables. Thus, although our conceptions of plural reference are similar, we identify significantly different expressions as plural referring ones, and in consequence what we take to be the implications of plural reference for logic and semantics are radically dissimilar.

Another author who does not try to reduce plural referring expressions to other kinds of expression is Peter van Inwagen, in his *Material Beings* (1990). Van Inwagen counts, as plural referring expressions, plural definite descriptions and conjunctions (but not disjunctions) of proper names and definite descriptions. He never explains what he means by plural reference, since, he claims, 'the idea of "plural referring expression" has sufficient currency' (p. 23). Yet in his note to the quoted sentence (note 7, p. 286) he refers only to Black's 1971 paper and to a paper by Adam Morton (1975), whose conception of plural reference is *reductive*.

Van Inwagen's discussion of plural reference (pp. 23-8) is far removed from mine. Firstly, like Black and Simons, he does not count common nouns among plural referring expressions. Secondly, he uses plural referring expressions *only* with variably polyadic predicates, i.e., 'predicates containing free plural variables' (pp. 23, 28), that usually express multigrade relations (for the exceptions, see van

Inwagen's reservations on page 28). Van Inwagen does not explain what a multigrade relation is, but his examples and his mentioned reference to Morton's paper make that clear. These are relations expressed by predicates like 'are in a minority', 'are quarreling' or 'are carrying a beam', which do not take any fixed number of arguments; the subject of 'are in a minority' can be a list of names of any length (Morton, 1975, p. 309; the concept and term are derived from Goodman and Leonard, 1940). By contrast, I think that most predicates can be used with *both* singular and plural referring expressions; 'asleep', for instance, should not appear with a plural referring expression according to van Inwagen, but that is what happens, I claim, in 'The children are asleep'. Lastly, like Simons, van Inwagen tries to *enrich* the predicate calculus by adding to it plural variables and existential and universal plural-quantifiers (pp. 25-6). By contrast, I shall rely on my claim that common nouns in quantified noun phrases are referring expressions to supply an *alternative* analysis of quantification in natural language.[2]

The last author who has written on plural reference and whom I shall discuss in this section is Byeong-uk Yi. In his paper 'Is Two a Property?' (1999) Yi argues, like Black, Simons and others before him, and against the standard conception, that many things as such can instantiate a property. He gives examples of two kinds of such an instantiation. The first involves multigrade properties and relations – e.g., 'live together' – which, like van Inwagen and unlike Morton, Yi construes non-reductively. The second involves numerical properties – e.g., 'Socrates and Plato are two'. Yi, like Simons, thinks that *being two* is a property, instantiated by many things as such (I shall discuss this view shortly). Yi develops a logic of plurals, very similar to van Inwagen's, 'which extends elementary logic to do justice to logical relations involving plural constructions' (p. 177). The expressions he mentions as plural referring ones are those mentioned by van Inwagen; nowhere does he mention the idea that common nouns function occasionally as plural referring expressions.

Yi's conception of plural reference is not entirely clear to me. Although his position, as I have just described it, would lead one to think that Yi's idea of plural reference should be similar to Black's, Yi writes the following on plural reference, which I do not know how to interpret (p. 176):

The plural term 'Bill and Hillary' ... refers to some things, namely, Bill and Hillary (as such). This does not mean that the plural term refers to Bill and also to Hillary; it refers to neither of them. A typical plural term refers to some things without referring to any one of them.

---

[2] Adam Morton also gives 'a formal account of multigrade relations and some related idioms' (1975, p. 309) that is based on the first order predicate calculus. In contrast to van Inwagen, Morton gives a *reductive* analysis of plural referring expressions: he analyzes plural definite descriptions as predicates, and predication on a conjunction of *n* names as predication by a multigrade predicate with *n* arguments (pp. 309-11). His purpose, translating the calculus of individuals into his notation, is very different from mine, and his analysis of natural language may have served him only as a point of departure.

I cannot see what referring to Bill and Hillary without referring to Bill could mean; rather, it seems to me that 'Bill and Hillary' refers to Bill and Hillary precisely because it 'refers to Bill and also to Hillary'.

In addition, Yi distinguishes (p. 187) between one-place predicates and their plural expansions, which he writes as, for instance, 'is-a-human' and 'is-a-human^P', respectively. The latter predicate indicates, according to Yi, a property which is the plural expansion of its singular base, a base that is indicated by the former predicate. But the property ascribed to Plato and to Plato and Socrates in 'Plato is human' and 'Plato and Socrates are human' is one and the same, nor is the predicate expanded or changed in the passage from a singular to a plural subject.

As I said, Yi thinks that *being two*, for instance, is a plural property. Similarly, Simons maintains that 'number is a property of external things of a kind which I call *manifolds*' (p. 161). Now I do think that the ascription of number, as in

> Plato and Socrates are two philosophers,

involves plural reference. But this does not entail that 'two' in that sentence is used to ascribe a property to Plato and Socrates. Yi thinks it does because he reduces such sentences to conjunctions – in this case, 'Plato and Socrates are two and Plato and Socrates are philosophers' (pp. 171-2). He therefore does not distinguish in his analysis between *being two* and *cooperate*, both of which he considers plural properties. But the reduction he proposes is implausible. Firstly, 'Plato and Socrates are two' feels like an ellipsis: it seems that a count noun – for instance, 'philosopher' – must at least be implicit. Secondly, such a reduction would not work with an example such as

> Plato, Socrates and Thucydides are two philosophers and one historian.

This fact is difficult to account for on Yi's approach, since this last sentence should be, according to him, semantically similar to 'Plato, Socrates and Thucydides are intelligent Athenians and impressive Greeks', which *is* reducible to the appropriate conjunction. Thirdly – and this objection applies to Simons as well – it would seem strained to claim that 'a few' is used to ascribe a property in, for instance

> Aristotle, Plato and Socrates are a few philosophers.

And similarly for other quantifiers. Simons and Yi should in this case justify why only some quantifiers ascribe properties in such uses. Lastly, 'two' is not used to ascribe a property when it is used as a determiner in a noun phrase used as a subject term – e.g., 'Two philosophers are Athenians'. And this is true of other, non-numerical quantifiers as well. So Simons and Yi should account for this dual use of numerical quantifiers according to their analysis, a dual use which is not paralleled in the case of other predicates.

I therefore think that it is more plausible to classify 'two' and other number-words as quantifiers, when they are used as determiners either in the subject or in the predicate, and accordingly, following the Medieval tradition, to consider them syncategorematic terms. This will be my approach later in this book (§ 0).[3]

Although the concept of plural reference seems straightforward and innocuous, it is still possible to maintain that on closer inspection it will reveal some logical incoherence, at least if used in tandem with some expressions of other semantic kinds. This indeed seems to be Strawson's position with respect to plural referring expressions formed by a conjunction or disjunction of singular referring expressions. Strawson argued (1974, chap. 1, § 1.2) that if we admit conjunction and disjunction of predicates ('tall and handsome', 'tall or handsome') in logic, we cannot coherently admit conjunction and disjunction of names of particulars ('John and Mary', 'John or Mary'). Admitting both, he argued, would enable us to make some invalid inferences.

For instance, from the premise 'John or Mary is tall and John or Mary is handsome' we could derive the sentence 'John or Mary is tall and handsome'; and from that sentence we could derive the conclusion 'John is tall and handsome or Mary is tall and handsome'. But while the premise is true in case John is tall while Mary is handsome, the conclusion is then false. A parallel argument can be constructed with subject-conjunction and predicate-disjunction.

As I said, Strawson concluded from this that we cannot coherently admit both subject combination and predicate combination in logic. He further argued that only the latter should be admitted in logic. If that were the case, then at least some kind of plural referring expressions would be logically unacceptable. It is unclear to me what such unacceptability would imply concerning plural reference in natural language. But whatever these implications might be, I shall try to show that the incoherence Strawson noted is avoidable even if both subject- and predicate-combination are admitted in logic.

To avoid the mentioned invalid inferences, we should define a sentence with predicate conjunction (or disjunction) as equivalent to a conjunction (or disjunction) of sentences only in case the subject is a singular term. That is, 'John is tall and John is handsome' would be defined as equivalent to 'John is tall and handsome'; but 'John and/or Mary are tall and John and/or Mary are handsome' should not be *defined* as equivalent to 'John and/or Mary are tall and handsome'. We can then define 'John is *F* and/or Mary is *F*' as equivalent to 'John and/or Mary are *F*', whether or not '*F*' is a compound predicate. Returning now to Strawson's argument, these definitions block the inference from the premise 'John or Mary is tall and John or Mary is handsome' to the sentence 'John or Mary is tall and handsome'. On the other hand, we can still use them to *prove* the equivalence of 'John and Mary are tall and John and Mary are handsome' to 'John and Mary are tall and handsome'.

---

[3] Two other philosophers who have favorably mentioned plural reference are David Armstrong (1978, pp. 32-4) and David Lewis (1991, pp. 62-71). They each derived the idea from some of the works I discuss in my book, and did not develop it beyond what can be found there.

Strawson himself maintained (ibid., p. 8) that the English sentence 'Either Tom or William both rides and drinks' is read in accordance with the definitions I suggested, and that in its case the invalid inferences are inadmissible in English. These definitions are therefore not ad hoc. They may, however, impose on natural language more logical order than it exhibits: perhaps some sentences that involve both subject- and predicate-combination *are* ambiguous. (Although statements made with them would usually be disambiguated by their context.) In any case, the use of such plural referring expressions surely exists in natural language, and at least much of it can be maintained without entailing any logical incoherence.

Strawson's ultimate reason for rejecting subject combination and preferring predicate combination is his thinking of a conjunctive or disjunctive subject term as designating, respectively, a conjunctive or disjunctive particular (1974, pp. 27-9; cf. his 1970, pp. 109-10). As if when we say 'John and Mary are tall' we refer to a particular which is the union of John and Mary and say that it is tall. Such particulars are clearly problematic, but they are certainly not what a conjunctive subject designates. When we say that John and Mary are tall we refer to two particulars, and we refer to them as *two* particulars, and say that each of them is tall. No bizarre conjunctive (or disjunctive) particular is intended. Accordingly, Strawson's metaphysical reason for rejecting subject combination is also flawed.

My conclusion in this section is, therefore, that the idea of plural reference is straightforward, and that various expressions in natural language seem to be frequently used as plural referring expressions. These include plural pronouns, demonstratives and definite descriptions, and conjunction or disjunction of singular and plural referring expressions. In fact, I believe that once the semantic category of plural referring expressions is pointed out, it seems so natural that its absence from a system of logic and semantics which purports to analyze those of natural language strikes one as unjustified.

We should now proceed to examine why this has generally been denied by philosophers and linguists, and whether their reductive analyses have been successful.

2.3   The Implausibility of Reductive Analyses of Plural Referring Expressions

In this section I shall argue that the attempts to reduce plural referring expressions to expressions of other kinds are unjustified by linguistic phenomena, and that the sole motivation for these attempts seems to be the presupposition that the functioning of these expressions *must* be analyzable by the semantic resources of the predicate calculus. I shall also try to show that the reductions that have been suggested are either mistaken or implausible.

Philosophers and linguists have suggested various paraphrases of sentences in which there are apparent plural referring expressions, paraphrases that contain only expressions belonging to semantic kinds that the predicate calculus recognizes –

singular referring expressions, predicates, connectives and quantifiers. It is indeed possible that some of these paraphrases have the same truth-conditions as the sentences they paraphrase. Still, this would not entail that the paraphrases reveal the way the expressions function in the paraphrased sentences. As Wittgenstein has remarked (*Philosophical Investigations*, § 22):

> We might very well also write every statement in the form of a question followed by a "Yes"; for instance: "Is it raining? Yes!" Would this show that every statement contained a question?

Similarly, '*p&q*' and '*-(-pV-q)*' are logically equivalent, yet the first sentence does not contain, in a covert form, negation and disjunction, nor does the second implicitly contain conjunction. Likewise, '*p*' and '*p&(qV-q)*' have the same truth-conditions, yet the former is not synonymous with latter.

The question we shall be asking is, what are the words doing in the analyzed or paraphrased sentence, the one containing apparent plural referring expressions? I shall try to show that even when the paraphrase may have the same truth-conditions as the paraphrased sentence, it is implausible that it shows that what seemed to be a plural referring expression is actually an expression of a different kind. The nature of the implausibility will consist in two things. Firstly, in the assumption of a large gap between what we take ourselves to mean, as this is revealed by our common or garden explanations, and what we are supposed to mean according to the paraphrase. Secondly, in the use in the paraphrase of theoretical concepts – such as those of a set and of membership in a set – which are supposed to be used by anyone who uses the paraphrased sentence, including those who have not been taught these theoretical concepts. If we were trying to devise a language with an expressive power resembling that of natural language, but having a narrower range of semantic kinds of expression, then such paraphrases would serve our purpose. But our project is different: understanding the way certain expressions contribute to the meaning of sentences.

I understand by a *semantically isomorphic translation* a translation which translates every expression or feature with a specific semantic function by an expression or feature with the same semantic function. If plural referring expressions are irreducible to expressions of other semantic kinds, then the predicate calculus is incapable of supplying semantically isomorphic translations of sentences of natural language that use the former expressions.

Let us examine the sentence

1   Paul is asleep.

It is translated into the predicate calculus as

2   *Pa*

where '*a*' translates 'Paul' and '*P*' 'is asleep'. 'Paul' is taken to be a referring expression, and is translated by the referring expression '*a*'. Although the speaker would usually identify Paul by some of his properties (including the way he looks), 'Paul' is not a predicate attributing these properties but an expression used to refer to Paul. And even if one maintains that a proper name has something like what Frege called '*Sinn*', 'Paul' does not designate a *Sinn* but Paul.

Similarly, the sentence

3   He is asleep,

uttered, say, while pointing at Paul or as an answer to the question 'Where is Paul?', is also translated as '*Pa*', where '*a*' translates 'he', which is taken to be a referring expression, used on this occasion to refer to Paul. Perhaps, as has been suggested by some, singular terms of a different kind should be introduced into the calculus in order to translate pronouns, demonstratives and indexicals, so that they will be distinguished from proper names; but the function of these terms would still be to refer. And as with the name 'Paul', the fact that the speaker would usually identify the person whom the pronoun designates by certain of his properties, does not turn this pronoun into a predicate attributing these properties.

Let us now look at the sentence

4   They are asleep,

uttered, say, while pointing at an unknown number of people sleeping in a room. The speaker uses the word 'they' to refer to these people – 'they' is a referring expression. Again, the speaker would usually identify the people referred to by means of some of their properties; but as with 'Paul' and 'he' above, this is no reason to consider 'they' a predicate attributing these properties. The only important semantic distinction between 'they' and the singular expressions 'Paul' and 'he' is that 'they' is a *plural* referring expression, denoting several particulars. If language and its use justify considering 'Paul' and 'he' referring expressions, then they justify considering 'they' a referring expression too.

Can one translate (4) into the predicate calculus? Since the calculus does not contain plural referring expressions, one would have to substitute other expressions for 'they'. Should one substitute a *predicate* for it, say '*Q*', and then translate (4) as, for instance:

5   (every x)($Q x \rightarrow$ Asleep x)?

Even if a predicate that yields the same truth-conditions can be found, this will not justify the claim that 'they' in sentence (4) is actually this predicate, or part of a construction in which this predicate is used, albeit in an opaque way. Since such a claim was not justified when translating (1) and (3), it is not justified here either. Moreover, no predicate is explicit in sentence (4), so the meaning of a claim that a predicate is in some sense implicitly contained in it is unclear. Of course, the speaker may be thinking all sorts of things – as he or she have done while uttering sentences (1) and (3) – but the question is, what does the *sentence* contain.

I believe it is clear that no sentence of the form of (5) is semantically isomorphic to sentence (4). And no other sentence of the predicate calculus is: the predicate calculus cannot translate sentence (4) by a semantically isomorphic translation because it lacks plural referring expressions.

I shall next discuss sentences in which singular and plural definite descriptions are used in the subject position. For instance:

6   My child is asleep.
7   My children are asleep.

In his paper 'On Denoting' of 1905, Russell maintained (p. 488) that sentence (6) means the same as

8   One and only one entity is a Child of mine, and that one is asleep.

In sentence (8), 'Child of mine' is a predicate, and this sentence can be translated into the predicate calculus while preserving the predicative role of 'child of mine' by, for instance,

(There is an x)((every y)(Child-of-mine $y \leftrightarrow y = x$) & Asleep x).

(A closer approximation to the syntax of (6) can be achieved by introducing binary quantifiers; see section 6.4 below.) The early and wide acceptance of Russell's analysis made it natural to regard sentence (7) too as an implicitly quantified sentence, in which 'Child-of-mine' is a predicate, and to translate it on the lines suggested by (8) above. For instance:

(every x)(Child-of-mine $x \rightarrow$ Asleep x).[4]

Russell's analysis was later very strongly criticized by Strawson (1950). Contrary to Russell, Strawson considered many of the definite descriptions Russell discussed to be referring expressions. Strawson's criticism and position were accepted by many, yet many logicians still think that Russell's analysis is at least approximately true. *In this work I adopt Strawson's position.* That is, I shall assume below that singular definite descriptions, in some of their uses, are referring expressions, and not quantified constructions or 'incomplete symbols' of the kind Russell took them to be. I shall not defend this position here, though, since such a defense would constitute a

---

[4] Cf. Neale, 1990, pp. 45-6. We would get an even closer relationship between the singular and plural case if we translate Russell's paraphrase (8) in the equivalent form

(There is an x)(Ch. x) & (every x)(every y)(Ch. x & Ch. $y \rightarrow y = x$) & (every x)(Ch. $x \rightarrow$ Asleep x),

which is a conjunction of an existence condition, uniqueness condition, and predication. We can then translate sentence (7) as

(There is an x)(Ch. x) & (every x)(Ch. $x \rightarrow$ Asleep x),

where the uniqueness condition, specified by the *number* of the definite description, has been omitted.

long digression. I believe that the fact that Strawson's position is recognized as a defensible option in contemporary semantics entitles me to do that.

I shall nevertheless note that even if the criticisms leveled against Russell by Strawson and others can be met,[5] the original puzzles that drove Russell into his *prima facie* implausible analysis were resolved by Strawson (1950, 1964). The only motivation left for adopting Russell's analysis thus seems to be that it is the only analysis available if one is limited to the semantic categories of the canonical version of the predicate calculus. In this book I provide several reasons for considering the calculus impoverished in its semantic categories compared with natural language. If I am right, it would be unreasonable to insist on a non-intuitive analysis of definite descriptions only because that is probably all that the calculus can offer.

If we adopt Strawson's position with regard to the use of 'my child' in sentence (6), 'My child is asleep', then it seems that we should also consider the use of 'my children' in (7), 'My children are asleep', referential. The function of the latter expression in language and communication parallels that of the former, apart from the fact that it is used to make an assertion about more than a single individual. It is clear that any attempt to translate (7), but not (6), into the calculus by *eliminating* this plural reference would be motivated only by the absence of plural referring expressions from the calculus.

Returning to sentence (6), 'My child is asleep', if we take 'my child' to be a referring expression, then we cannot translate it into the canonical version of the predicate calculus while preserving that expression's function. The canonical version of the calculus has no referring expression to translate natural language's referring expression 'my child'. This lack can be accommodated, however, by introducing Russell's inverted iota operator (see also Frege, *Grundgesetze der Arithmetik* I, § 11). Let '$(\iota x)Fx$' denote the only thing which is F, if there is such a thing, and nothing otherwise. Then (6) can be translated as

Asleep$((\iota x)$My-child $x)$.

This enrichment of the calculus is relatively straightforward. The calculus already has a slot for singular referring expressions, and therefore all that is needed is the invention of a new kind of singular referring expression. This translation indeed makes the use of 'my child' as a referring expression involve its use as a predicate, while this may seem not to be the case in (6). But this is a separate difficulty, while the problem of introducing singular referring expressions that have some content *is* solved by Russell's iota operator.

Parallel considerations should bring us to introduce a *plural* inverted iota operator into the calculus, if we want the calculus to translate sentence (7) by a semantically isomorphic sentence. Such a move would concede my point that the plural definite description in (7) is a plural referring expression, and that expressions of this kind do not exist in the calculus.

---

[5] The best defense of a Russellian analysis I know is Neale's, in his *Descriptions*.

However, as I shall show below (Chapter 6), the absence of plural referring expressions from the calculus necessitated its departure from natural language in its treatment of quantification, and this departure cannot be rectified by adding any new kind of expression.

Let us next consider sentences in which the grammatical subject is formed by a conjunction of singular referring expressions. For instance:

9   Tom and Jane went to sleep.
10   Tom and Jane mowed the whole meadow.

Sentence (9) is true if and only if its predicate applies to each of the particulars its conjuncts denote; i.e., Tom and Jane went to sleep if and only if Tom went to sleep and Jane went to sleep. By contrast, according to the most salient reading of sentence (10), Tom and Jane mowed the whole meadow only if neither Tom mowed the whole meadow nor did Jane. Predication as in (9) is called *distributive*, while that in (10) is called *collective*.

If we consider 'Tom and Jane' in both (9) and (10) plural referring expressions, designating both Tom and Jane, then the semantic structure of both (9) and (10) is clear. In both sentences two persons have been referred to and a predicate has been used. Since 'went to sleep' is distributive and the names in (9) are connected by a conjunction, the predicate should apply to both persons mentioned. (If the names were connected by a disjunction, the predicate should apply to at least one; cf. above, p. 9.) While since 'mowed the whole meadow' is used collectively in the reading of (10) we now consider, it should apply to both persons mentioned *together*. (But here too, if the names were connected by a disjunction, the predicate would have to apply to at least one of the persons to which the names refer.)

However, if one believes that the semantics of both (9) and (10) should be analyzable by the semantic resources of the predicate calculus, then the absence of plural referring expressions from the calculus should make one attempt to reduce the apparent plural referring expression 'Tom and Jane' to other kinds of expression.

That was indeed Frege's approach. In a letter to Frege dated 10 July 1902, Russell maintained that 'classes cannot always be admitted as proper names. A class consisting of more than one object is in the first place not *one* object but many (*viele*).' (Russell does not distinguish here between a class and a class name, a distinction which Frege will make in his reply.) Similarly, Russell writes further on that 'certain classes are mere manifolds (*Vielheiten*) and do not form wholes (*Ganzes*) at all.' Russell thought that the distinction between a class as one object and a class as many objects can help resolve the famous contradiction he discovered in Frege's logic, a contradiction which he pointed out to Frege in his letter from 16 June of that year. Russell relied on this distinction in his discussion and rejection of this contradiction in his *The Principles of Mathematics* of 1903 (section 70 and chapter X).

Russell's position, which is similar to mine on plural referring expressions, was unacceptable from Frege's point of view (and as is clear from Russell's letter from 8

August 1902, Frege succeeded in convincing Russell to reject this position[6]). In his reply to Russell (dated 28 July 1902) he writes that 'if a class name is not meaningless, then, in my opinion, it means an object. In saying something about a manifold or set (*Menge*), we treat it as an object.' He then proceeds to distinguish three cases, two of which are of sentences containing conjunctions of proper names. In Frege's analyses of these apparent plural referring expressions, expressions referring to 'many objects', the plural reference is of course eliminated. These analyses reappear in his posthumously published 'Logic in Mathematics', written in 1914.

Where the predication is what I called *distributive*, Frege maintained that 'we are not really connecting the proper names by "and"', but telescoping two connected sentences into one (1914, p. 227); his examples being 'Schiller and Goethe are poets' (ibid.) and 'Socrates and Plato are philosophers' (Letter to Russell; cf. also his 1884, note to section 70). That is, 'Tom and Jane went to sleep' is, according to Frege, really a conjunction of two sentences, 'Tom went to sleep and Jane went to sleep', contracted into a single sentence for the sake convenience (Letter to Russell). By contrast, where the predication is what I called *collective*, as in 'Bunsen and Kirchhoff laid the foundations of spectral analysis', Frege claimed that we consider Bunsen and Kirchhoff as one whole or as a system, as we do a nation, an army or a physical body (Letter to Russell; cf. 'Logic in Mathematics' p. 227-8, where the reference is said to be to a 'compound object'). In the distributive case the 'and' connects sentences, while in the collective case it is 'used to help form the sign' for a compound object (1914, pp. 227-8). Frege in fact distinguishes two kinds of *reference*, and not of predication, of conjoined names. While reference in (9) is repeated singular reference to ordinary particulars in logically separate propositions, that in (10) is singular reference to a system or a compound object. Analyses similar to Frege's were suggested time and again during the last century, occasionally perhaps independently of his.

Frege's analyses should be rejected for several reasons. Consider, first, the sentence

11   Tom and Jane mowed the whole meadow and went to sleep.

On Frege's analyses, 'Tom and Jane' in (11) is ambiguous: on the one hand, with respect to the predicate 'mowed the whole meadow', it refers to a system or a compound individual, and 'and' is 'used to help form the sign'; on the other hand, with respect to the predicate 'went to sleep', it is an abbreviation of sentence conjunction, and 'and' signifies that conjunction. But this is implausible: we do not feel any change of meaning of the subject term when we pass from the first predicate to the second one. By contrast, no such ambiguity is involved if we consider 'Tom and Jane' a plural referring expression.

---

[6] Russell writes there that he now understands 'the necessity of treating ranges of values not merely as aggregates of objects or as systems.'

Modern linguists, influenced by model-theoretic semantics, followed Frege in considering the distinction between distributive and collective to consist in the manner of reference. The idea that the distinction is in the manner of predication is often missing from their discussions (see Lønning, 1997). In consequence, sentences like (11), involving both collective and distributive predication, generate spurious difficulties for them (ibid., § 5.1). The situation is similar in philosophy, where the mistake of considering reference, and not predication, either collective or distributive has driven philosophers to unnecessary ontological elaboration (cf. Cameron, 1999, p. 129). (Oliver and Smiley (2001) are the only exception I know of; for reasons similar to mine, they consider predication, and not reference, either distributive or collective (pp. 292-5).)

Secondly, the comparison of Bunsen and Kirchhoff, even in this limited respect, to a nation, army or body is surely odd. And considering Tom and Jane as forming together some kind of complex or structured individual is no less bizarre. *Pace* Frege, we do not refer to either pair as to one composite thing, in the way that we refer to a nation or an army – collective bodies that can preserve their identities even if some or all of their members are changed. We refer, say, to Tom and Jane as two people, who did something together. Perhaps each one mowed half the meadow, perhaps they pushed the mower together, or they may have cooperated in some other way; at the end the whole meadow was mowed. The claim that sentences with collective predication involve singular reference to wholes or to compound objects seems implausible and ad hoc.

Lastly, although sentence (9), 'Tom and Jane went to sleep' describes the same situation as 'Tom went to sleep and Jane went to sleep', and although each sentence entails the other, this still does not justify considering the former an explicit conjunction of two sentences. As I have said above, '$p\&q$' and '$-(-pV-q)$' are logically equivalent, yet the first sentence does not contain, in a covert form, negation and disjunction, nor does the second implicitly contain conjunction. An independent reason should be supplied for considering the paraphrase as revealing the real nature – in some sense of 'real' – of the paraphrased. And the fact that the claimed equivalence of names- and sentence-conjunction is not general, as is demonstrated by (10), weakens the plausibility of the claim that (9) is actually a conjunction of sentences. It rather seems that the use of 'and' as a connective of names and its use as a connective of sentences are sometimes equivalent, and therefore the same word is used – not only in English – in both cases; while these uses are frequently nonequivalent, and therefore none can be reduced to the other.

Frege's analysis of sentence (10) is thus implausible, and his analyses of both (9) and (10) seem to be motivated not by any linguistic phenomenon, but rather by the absence of the appropriate expressions from his calculus. His analyses should therefore be rejected.

Following Donald Davidson's influential analysis of action sentences (1967), philosophers and linguists tried to extend his analysis to sentences with plural nouns as subjects. Already Castañeda, in his commentary on Davidson's paper (1967), suggested to analyze the sentence

>   Anthony and Bill (making up a team) won

as either

>   (There is an event e)(Won(Anthony-Bill, e))

'where the hyphen indicates a Goodman-type of summation of individuals,' or

>   (There is an event e)(Won({Anthony, Bill}, e))

'where the braces indicate the set whose members are listed or described within them' (p. 107). But, as Oliver and Smiley argue (2001, p. 299), 'neither of Castañeda's candidates for agents will do. Sets are abstract and so cannot win things.' And to consider Tom and Jane some kind of complex or structured individual just because they mowed the meadow together is surely unjustified, as I argued above against Frege.[7]

James McCawley (1968, pp. 152-3) suggested a different variation on the Davidsonian theme as an analysis of sentences like (10), 'Tom and Jane mowed the whole meadow', where the predication is collective. In contrast to Castañeda, McCawley does not see the agent in the event as some kind of a compound object, but he considers every member of a set as an agent of the event. Oliver and Smiley (2001, § IV) slightly elaborate McCawley's analysis. They would analyze sentence (10) as follows:

>   There is an event which is a mowing of the whole meadow and every member
>   of {Tom, Jane} played a part in it and no one else did.

On this analysis, '{Tom, Jane}' is a singular term referring to a set; in this way the plural reference to Tom and Jane is replaced by a singular reference to a set. And Tom and Jane are those who mowed the meadow, and not any abstract or compound object.

Oliver and Smiley do not explicitly endorse or reject this analysis as a correct one for specific cases. They do ultimately reject it as a *general method* of paraphrasing plural action sentences that avoids plural reference, since they argue that as a general method it would generate Russell-like paradoxes (ibid., § V). But I shall try to show that this analysis is generally implausible.

Firstly, this analysis is clearly motivated only by the desire to eliminate plural reference. Even if one adopts the general Davidsonian analysis of action sentences, then, if one allows of plural reference, sentence (10) can be analyzed as follows:

>   There is an event which is a mowing of the whole meadow and Tom and Jane
>   did it.

---

[7] I think Oliver and Smiley's own argument against considering Anthony and Bill 'a Goodman-type of summation of individuals' (ibid.) unsound, and therefore I do not use it here.

Secondly, the claim that sentence (10) implicitly contains reference to a set, membership in a set and quantification over its members is surely counter-intuitive. Thirdly, consistency would compel Oliver and Smiley to translate the sentence 'Tom mowed the meadow' as 'There is an event which is a mowing of the meadow and every member of {Tom} played a part in it and no one else did'; and this time the unproblematic singular reference to a person is substituted by reference to a set whose only member is that person.[8] Fourthly, consider sentences with conjoined collective and distributive predication, as in (11), 'Tom and Jane mowed the whole meadow and went to sleep.' Here one should not analyze (11) as containing a single existential quantification over events, since mowing the whole meadow seems to be a different event from going to sleep. But then, 'Tom and Jane went to sleep' should not be analyzed as containing a single existential quantification either, since they need not have gone to sleep together. So it should be analyzed as a telescoping of two sentences, 'Tom went to sleep and Jane went to sleep.' And as in the case of Frege's analyses, we end with an alleged ambiguity of 'Tom and Jane' where none is apparent.

Lastly, this analysis eliminates plural reference only apparently. Language should be able, presumably, to specify which set is the set referred to. In the case of the set {Tom, Jane}, this can be done by saying that it is the set that has Tom as a member and Jane as a member, and nothing else – here plural reference is indeed eliminated. But suppose a plural pronoun, demonstrative or definite description is used, for instance:

We mowed the whole meadow.

What are the members of the set {We}, a set that will presumably be used in Oliver and Smiley's analysis? It is the set that has *us* and only *us* as members – and here we have plural reference again. Oliver and Smiley's analysis relocates plural reference, but it does not eliminate it. (They themselves bring a similar objection against a different analysis in the penultimate paragraph on page 297.)

Oliver and Smiley's analysis should therefore also be rejected as implausible. And even apart from its implausibility, it does not succeed in showing that plural reference is reducible to other semantic kinds.

A different variation on the Davidsonian theme was attempted by Schein.[9] Schein analyzes sentences involving plurals as having 'a logical form that derives from

---

[8] Already Castañeda acknowledged this consequence (1967, p. 107). It seems, however, to have escaped Oliver and Smiley, who analyze the singular case without reference to the appropriate set (2001, pp. 299-300).

[9] First in his unpublished 1986 doctoral dissertation, then by Higginbotham and Schein in a paper called 'Plurals' (*Proceedings of the North Eastern Linguistics Society* 19: 161-75), and most elaborately in Schein's 1993 book, *Plurals and Events*. My account of Schein's theory is based on his book. Higginbotham has subsequently criticized and rejected Schein's analysis, in favor of a Russelian conception of classes as many, similar in certain respects to my conception of plural reference (Higginbotham, 1998, §§ 7-8).

Donald Davidson 1967 ... as emended by Castañeda (1967)' (Schein, 1993, p. 3 and note 6), while also applying Boolos's second-order predicate calculus in his analysis. (I discuss Boolos's logic below, p. 85.) According to Schein's analysis, the logical form of the sentence

   The elms are clustered in the forest

is transparently given by the following translation into the second order predicate calculus (where 'INFL' is 'the relation between *each* elm and the event of being clustered') (ibid., pp. 3-4):

$$\forall x(\text{INFL}(e, x) \leftrightarrow (\iota Y)(\exists y Yy \ \& \ \forall y(Yy \leftrightarrow \text{elm}(y)))(x)) \ \& \ \text{cluster}(e) \ \& \ \text{In}(e,$$
the forest).

This sentence roughly says that every thing is among the subjects of the event if and only if it is one of the elms, and the event is clustered and in the forest; while 'the elms' is construed as 'the Y such that at least something is a Y and anything is a Y if and only if it is an elm' – 'Y' is here a predicate, as can be seen.

I believe the first impression this analysis makes is of being convoluted beyond acceptability. And Schein's analysis faces additional criticisms as well.

Firstly, it involves, as Schein acknowledges, the assumption that 'all predicates are, first of all, about events' (ibid., p. 3; cf. Higginbotham, 1998, p. 20), and this seems implausible. An event is something that happens, but many predicates do not require reference to anything of the sort. What could be the event referred to in 'John is intelligent'? To construe it as involving reference to an event in the same sense that one is involved – *if* one is involved – in 'John sneezed at 12:00 o'clock', is to abandon the sense of 'event' for the sake of some dummy index of some syntactical significance within the framework of some formalism. Even the claim that reference to a specific event is involved in the sentence 'The elms are clustered in the forest', used by Schein as his basic example, is implausible.

Secondly, Schein's approach commits him to construing proper names as predicates, especially given his claim that 'second-order logic is the logic for all plurals' (p. 39). In order to analyze sentence (10), 'Tom and Jane mowed the whole meadow', Schein would have to construe 'Tom and Jane' as a predicate, true only of Tom and of Jane. His approach also commits him to construing plural demonstratives as predicates, e.g. 'they' in 'They built a boat yesterday' (Higginbotham, 1998, p. 23). But if we do not take 'Tom' and 'he' to be predicative, there is no linguistic justification for taking 'Tom and Jane' or 'they' to be such. On the other hand, if Schein would not apply his analysis in these cases, then he would both introduce a distinction in logical form where none is apparent, and an important group of plural referring expressions would be left unanalyzed.

Thirdly, as in Oliver and Smiley's case, Schein would be compelled to claim that the use of a plural noun phrase as the subject of a sentence with both collective and distributive predication involves ambiguity, which is implausible (see the fourth objection to Oliver and Smiley above).[10]

Consequently, like the former reductive analyses of plural reference that we have considered, Schein's should also be rejected, for being implausible in several respects.

Thus, my conclusions in this section are that, firstly, there is no linguistic phenomenon that justifies a reductive analysis of plural referring expressions, and, secondly, that the reductive analyses suggested are at least implausible. We should take the expressions discussed in this and the preceding section for what they appear to be: expressions referring to more than a single particular.

---

[10] Higginbotham points out yet another difficulty in Schein's analysis – see Higginbotham, 1998, pp. 22-3.

Chapter 3

Common Nouns as Plural Referring Expressions

3.1   The Functioning of Common Nouns

My main claim in the previous chapter was that in contrast to the predicate calculus, natural language has not only singular, but also plural referring expressions. Expressions that are used in this way include, I argued, plural pronouns, demonstratives and definite descriptions, and conjunctions or disjunctions of these and of singular referring expressions. That is, I claimed that natural language has expressions belonging to a semantic category that is absent from the predicate calculus.

As far as this claim is concerned, one would only need to enrich the calculus in order to make it capable of supplying semantically isomorphic translations of natural language sentences. Moreover, my claim did not disagree with any of the semantic claims of Frege that were essential to the development of his calculus.

In this chapter, however, I make a more radical claim, which does disagree with some of these semantic claims of Frege's. I shall argue that in many cases, *common nouns in quantified noun phrases are not predicative, but plural referring expressions*. Frege, by contrast, maintained that they are predicative. As I noted above (§ 2.1), already in his *Begriffsschrift* (§ 12) he translated the subjects in the four Aristotelian quantified sentences as predicates, and several times in his later writings he argued for this analysis.

Frege's analysis was almost unanimously accepted by later philosophers and linguists. Dummett, referring to this analysis as it appears in section 47 of Frege's *Die Grundlagen der Arithmetik*, applauds it as follows (1991, p. 93):

> There is no such thing as a 'plurality', which is the misbegotten invention of a faulty logic: it is only as referring to a concept that a plural phrase can be understood ... But to say that it refers to a concept is to say that, under correct analysis, the phrase is seen to figure predicatively. Thus 'All whales are mammals', correctly analyzed, has the form 'If anything *is a whale*, it is a mammal' ... On this analysis, no one has subsequently found an improvement, the only plausible variation being that which would substitute, say, 'any organism' for 'anything'..., importing an explicit circumscription of the domain into the [quantification].

To challenge this analysis of Frege's is to challenge an uncontroversial analysis, which is at the heart of his logic.

I shall proceed with this delicate project as follows. In this section I shall try to show that there is a *prima facie* good case for taking some common nouns in quantified noun phrases to be plural referring expressions. Then, in the rest of this chapter, I shall support this analysis by showing that it explains various linguistic and semantic phenomena. I shall then proceed, in Chapter 4, to examine and reject Frege's reasons and arguments for his analysis of common nouns in quantified noun phrases as predicative, as well as some additional arguments mentioned by Russell. This will lead to a discussion of reference, in Chapter 5.

Although I do not think that my arguments for my analysis and against Frege's are conclusive – how often does one come across conclusive arguments in philosophy, or in any other field of knowledge? – I do think they show that my analysis is far better supported than his.

We start, accordingly, with *prima facie* reasons for considering common nouns in some quantified noun phrases as plural referring expressions.

Consider the following two sentences:

1   You were asleep.
2   One of you was asleep.

Assume 'you' in sentence (1) is used as a *plural* pronoun. I argued in the previous chapter that expressions like the plural pronoun 'you' in that sentence are plural referring expressions. Now there does not seem to be any obvious reason to claim that this expression has changed its function in sentence (2). Rather, it seems more plausible, at least *prima facie*, to maintain that 'you' is still used to refer to several people, and that we use the quantifier to specify that the predicate applies to one of them. The case is similar with lists of names, plural demonstratives and plural definite descriptions. For instance:

3   These children are asleep.
4   Some/Many/Two of these children are asleep.

Again, there does not seem to be any *prima facie* reason to claim that the expression 'these children', although a plural referring expression in sentence (3), has changed its function in sentence (4). We still refer to children by means of it, but this time we say that a certain number of them, specified by the quantifier, are asleep.[1]

If this is conceded, we can now substitute a common noun for either the plural pronoun in (2) or the plural demonstrative in (4). For instance:

5   Some/Many/Two children are asleep.

---

[1] Sentences like (2) and (4) seem to have escaped Simons' notice, who writes that in English, plural terms 'may not be preceded by ... quantifier phrases' (1982, p. 208).

Since the common noun 'children' occupies the same place in the sentence as the plural pronoun and the plural demonstrative, it is natural to attempt to ascribe it the same semantic function, namely plural reference. 'Children' in sentence (5) seems to be used to designate several children.

Indeed, in sentences (2) and (4) we had the preposition 'of' coming between the quantifier and the plural pronoun or the plural demonstrative, while 'of' was omitted in (5). But it is difficult to see why this should imply that the plural pronoun or the plural demonstrative have a semantic function different from that of the common noun. Moreover, not all quantifiers require a preposition when followed by a plural demonstrative or definite description; we say both 'All horses are animals' and 'All the people you invited are coming'. In addition, some quantifiers require the same preposition in both cases, 'a minority of' being one example. And lastly, the existence of a preposition following certain quantifiers or preceding certain nouns is language-dependent, and this does not seem to affect the adequacy of translation between languages. So for all these reasons, we should not ascribe any semantic significance to such presence or absence of a preposition.

The correspondence we noted between the use of common nouns and that of plural referring expressions in quantified noun phrases goes even further. Consider the sentence:

Many students will pass the exam, but some of them will inevitably fail.

The grammatical correspondence between the common noun 'students' and the anaphoric plural pronoun 'them', and the semantic parallels between the two conjuncts, make it plausible to maintain that since 'them' is used to refer to certain students, say those who take the mentioned exam, so does 'students'.

Lastly, the use of the determiners 'the' or 'these' with the common noun in a quantified noun phrase is often optional. We can say either 'Some children are asleep' or 'Some of the children are asleep'; 'All students will take the exam' or 'All the students will take the exam'; and so on. If we consider plural demonstratives and definite descriptions in these uses referential, then since the presence of the determiner does not affect what is being said, we have a good reason to consider common nouns, when used without any determiner following the quantifier, referential as well. Moreover, in some languages the use of a definite article with the common noun in quantified noun phrases is obligatory with some quantifiers. For instance, in Hebrew, if one uses 'all' or 'most' with a common noun, it must be preceded by the definite article. But this does not seem to affect the adequacy of translation of English sentences in which the definite article is not used into Hebrew. So if the plural definite noun phrase is a plural referring expression in Hebrew, we have a good reason to consider the common nouns in English noun phrases like 'all children' and 'most children' referential as well. And this can then be generalized to other common nouns in quantified noun phrases.

In all my examples in this section the alleged plural reference was to individuals that could often be counted or even named. One who says 'All the people you invited are coming' may often answer the question, 'Which people were invited?'

by naming them. It is not clear how such examples can be generalized to reference to future or past individuals, say, or to infinitely many particulars. This question will be discussed in Chapter 5.

Yet we have several *prima facie* good reasons to consider many common nouns, in many quantified noun phrases, plural referring expressions.[2] We shall now proceed to see how this analysis of common nouns can explain various linguistic phenomena.

3.2   On an Alleged Ambiguity of the Copula

The analysis of common nouns in sentences like 'All men are mortal' not as predicative but as referring expressions leads to a reassessment of a widely accepted claim about the ambiguity of the copula. Starting with Frege's *Begriffsschrift* (§ 9, p. 17), it has commonly been maintained that despite linguistic appearances, the copula, or the copulative structure, has different meanings in singular and quantified subject-predicate sentences. For instance, in the two sentences

1   Socrates is mortal
2   Socrates and Plato are mortal

the copulas are used to indicate predication, mortality being predicated of Socrates and Plato. By contrast, in the sentences

---

[2] One objection to my claim, that common nouns, which are often plural referring expressions in natural language, are not analyzed as such by the predicate calculus, may run as follows. According to model-theoretic semantics, predicates *are* referring expressions – they refer to classes of particulars. (More accurately, one-place predicates refer to classes of particulars, while binary, three-place, etc. predicates refer to classes of ordered pairs, triplets, etc. of particulars.) And reference to a class of particulars at least resembles plural reference to these particulars. Common nouns, being predicates, are therefore at least similar to plural referring expressions, according to model-theoretic semantics.

But this objection founders on the absence of a distinction between reference and predication. Natural language distinguishes between the two; in 'John is intelligent', 'John' is a referring expression and *not* a predicate, while the converse is true of 'intelligent' – 'intelligent' does not designate anything. If one wants to maintain that in model-theoretic semantics predicates are referring expressions, then one is left without predication.

A double confusion is contained in the basis of model-theoretic semantics, and as a result in the way in which the semantics of natural language is conceived. On the one hand, common nouns are analyzed as predicates. Consequently, since predicates in natural language are not referring expressions, this status is officially denied to common nouns too, and only singular terms are declared referring expressions. But on the other hand, since common nouns in natural language *are* referring expressions in many of their uses, it is natural to mistakenly take predicates generally, of which common nouns are now considered a representative part, to refer to classes. In consequence, all predicates, including adjectives and verbs, are sometimes said to refer. The concepts of reference and predication thus become quite muddled.

3   Every Greek is mortal
4   Some Greeks are mortal

no predication is involved, but rather a determination of relations between concepts. While according to some semantics the copula in (1) should be seen as expressing a relation synonymous with set-theory's membership, expressed by '$\in$', sentence (3) should be read as asserting the inclusion or subordination of one class or concept to another, symbolized by '$\subset$'. Frege elaborated on these distinctions in his 'Kritische Beleuchtung einiger Punkte in E. Schröders *Vorlesungen über die Algebra der Logik*' (see his third point at the end of that essay; cf. his second note in 'Über Begriff und Gegenstand'), and he also maintained that the lack of this distinction in Euler's diagrams makes them a lame analogue for logical relations (1895, p. 441).

It is difficult to accept this claim for the ambiguity of the copula, for several reasons. Firstly, consider the sentence

5   Socrates and some other philosophers are Athenians.

What should the meaning of the copula 'are' in this sentence be, according to the ambiguity claim? Since something is claimed about Socrates, it should indicate predication. On the other hand, since something is allegedly claimed about the relation between two concepts – being a philosopher and being an Athenian – it should also indicate some conceptual relation. So should the copula in sentence (5) have some third meaning, a composite of its two other meanings? Or is it ambiguous? Meanings and ambiguities threaten to proliferate. This seems implausible.

Secondly, if a certain grammatical construction is ambiguous in one language, the reappearance of the same ambiguity in a second language that is historically unrelated to the first would be difficult to explain. However, the same alleged ambiguity reappears in Hebrew, which is grammatically very different from English.

By contrast, if common nouns in the subject position in subject-predicate sentences are referring expressions, then the alleged ambiguity does not exist. In (1) to (4) we say of some particulars – one, two, or many – that they are mortal. We refer to particulars, and predicate something of them. The same applies to (5), where both Socrates and some other particular philosophers are classified as Athenians. In all these cases, the copula indicates predication. The analysis of these common nouns as referring expressions explains away the implausible ambiguity of the copula generated by their analysis as predicates. It is thus more reasonable to maintain that the predicate calculus, and not ordinary language, is logically misleading in this case.

3.3   Attributive and Predicative Adjectives

In section 3.1 I maintained that common nouns are often used not as predicates, but to refer to particulars. I thus distinguished between the use of a concept to refer to particulars and the use of a concept as a predicate, to say something about particulars referred to by other means. This distinction should make us look for related differences between concepts used in these different ways. Perhaps the constraints on a predicative concept are different from those on a referring one, and as a consequence some concepts can be used as predicates but not as referring expressions, or *vice versa*. And perhaps referring concepts are semantically similar to other referring expressions in ways in which predicative ones are not. The detection of such differences and similarities will support my claim in section 3.1, that it is mistaken to consider all concepts as semantically or logically predicative, as is done by the predicate calculus.

Which concepts can be used as predicates, but not as referring expressions? Many adjectives are used attributively (in the philosophical sense), i.e., what they attribute to a particular of which they are predicated depends on its classification.[3] A good lecture is not good in the same way as someone's eyesight may be good; a big mouse is not the size of a big elephant; and so on. And an animal can be big qua mouse, while not big qua animal. This is not a change of meaning, though: 'big' does not require several separate entries in the dictionary, one for mice, another for elephants, yet another for animals generally.

Attributive adjectives cannot easily be used as referring expressions. The property they attribute depends on the presence of a noun which classifies the particulars to which they attribute the property. Consequently, if they were used as referring expressions, without any other concept determining the property by which they are to pick out the particulars referred to, it would be indeterminate which property is supposed to determine their reference. Thus, if they are to be used to refer to particulars, a noun that specifies which kind of particulars are talked about should be presupposed.

Consider, for instance, the attributive adjective 'little'. If we want to use it as a referring expression, then, since it is an adjective, we should use the expression 'the little one'. But in contrast to sentences like 'Jane will arrive late', 'She will arrive late' or 'The little girl will arrive late', the sentence 'The little one will arrive late' is either colloquial – where the audience knows that the speaker is talking, say, of a specific girl; or it continues a conversation in which a classifying noun has already been used to specify several particulars, and 'the little one' is now used to refer to the little one among them – e.g., 'When should the girls arrive?' – 'All I know is that the little one will arrive late.'

This phenomenon is not peculiar to English. Unlike English, some languages – e.g. Hebrew, Italian and Latin – do not have clear grammatical criteria that determine whether a word is an adjective or a noun. Unlike English, where adjectives, unlike nouns, do not have a plural form, in these languages the words that translate English adjectives have plural form as well as gender, like those that translate English nouns. Moreover, while, in English, sentences with adjectives as subjects are ungrammatical – while one can say 'The woman will come late', one cannot say 'The little will come late' but has to say 'The little *one* will come late' – in Hebrew, Latin and Italian the

---

[3] The meaning of 'attributive' in philosophy, which I have just defined, is not its grammatical meaning; in grammar it means that the adjective modifies a noun in a noun phrase, like 'old' in 'the old man'. The philosophical use is due to Geach (1956).

corresponding sentences are grammatical. Whether a word is used in a specific sentence as a noun or as an adjective is determined by its place in the sentence. All the same, the concept 'little' in all those languages behaves similarly to its behavior in English. As in English, the sentences translating 'The little one will come late', like other sentences with attributive adjectives as subjects, are used either colloquially, or when the kind of particulars referred to has been previously specified or is clear from the context.[4]

On the other hand, some adjectives are used predicatively, i.e., the property they attribute is independent of the classification of the thing to which they attribute it. Such adjectives can be used as referring expressions. The things they will then denote are those that have the property that they attribute when used as adjectives.

'Square' is an example of such a predicative adjective. Independently of the classification of the thing of which it is predicated, it is used to describe a certain form. Accordingly, if used in order to refer to something, it would be clear to what it refers – something that has a square form. We thus find 'square' used either as an adjective, attributing properties – e.g., 'This table is square' or 'I like the square table' – or as a referring noun – e.g., 'The floor was tiled in squares of gray and white marble'. The case is similar with the adjective 'triangular', which has the corresponding noun form 'triangle'. Cf. 'The room is triangular' and 'Triangles have three sides'. And again, in some languages – Hebrew, for instance – the same word is used both with the meaning of the English adjective 'triangular' and of the English noun 'triangle'.

How do these linguistic phenomena support my analysis of common nouns as against Frege's? Frege claimed that a concept-word like 'elephant', say, in the sentence 'Some elephants are huge', is predicative – as is the concept-word 'huge'; they both mean or designate (*bedeuten*) concepts. Accordingly, he would translate this sentence into his calculus as

$$(\text{Some } x)(\text{Elephant } x \ \& \ \text{Huge } x).$$

But then, the sentence 'Some huge ones are elephants' should also have had a similar sense according to his analysis – again, two concept-words are used as *predicates*. Yet this sentence depends on the context in a way the former one, 'Some elephants are huge', does not. It seems that something in Frege's analysis is amiss.

On the other hand, if the concepts used in such quantified noun phrases are used *referentially*, as I maintain, then if a concept cannot be used in this way, or if its use in this way is constrained by context in a specific manner, we should expect such asymmetries. And indeed, 'huge', being attributive, cannot be used referentially apart from very specific contexts. My analysis can therefore easily explain a phenomenon which creates difficulties for Frege's.

---

[4] For Hebrew, see Barri (1978), especially sections 2.22 and 2.2.3.1.

3.4   Natural Kind Terms

My discussion of attributive adjectives was intended to show that some concepts have a use as predicates but not as referring expressions, or that their use as referring expressions is secondary in some sense. Can we find the corresponding phenomenon with predication, i.e., concepts that can be used as referring expressions but not as predicates, except in some secondary sense?

I think we cannot. There are, however, some concepts of a special kind, whose use as predicates will be distinctive in important respects. I shall now discuss them.

Different people can refer to the same things by the same word, even if they identify the things they refer to by means of different properties. All that is needed for common reference is that every speaker use identifying properties that pick out the same particulars as those picked out by the identifying properties used by any other speaker. Accordingly, if an expression is used to refer to particulars, different users can identify the particulars referred to by means of different properties and yet the expression will have the same meaning for all, since they all refer to the same particulars.

This is clearly the case with proper names. Different people may identify the same person by different properties. One may have identified Plato by the way he looked, another – a blind man – by his voice, and yet another as the author of certain dialogues. All the same, they were all referring to Plato when they used his name. The name 'Plato' had therefore the same meaning, or function, for all. For instance, if one said, 'Plato is now at the market place', the others would understand him.

The case with some common nouns is analogous. For instance, different people attribute different identifying properties to dolphins: one may think only of their shape and habitat, another about their inner organs as well, yet another about these and about their behavior. But as long as they would count the same particulars as dolphins (if they were sufficiently well informed) all mean the same by the word 'dolphin'. 'Some dolphins have just swam past the ship' will mean the same for all. Different people may also explain what dolphins are by reference to different, logically independent, properties of dolphins; but these different explanations will all be satisfactory if they are sufficient for a correct use of 'dolphins'.

Not all common nouns have semantic properties similar to 'dolphins'. Those that do may perhaps be classified as *natural kind terms*. Natural kinds have an inexhaustible number of distinctive properties, which are not logically derivable from each other (Mill, 1872, 1.VII.4, 4.VI.4; cf. Russell, 1948, p. 335). Because of these characteristics, for every natural kind there is always the possibility that it will be discovered to have some hitherto unknown property, which will then be used to identify members of this kind. This is in contrast to common nouns like 'student' or 'stone'.[5]

---

[5] My explanation of the meaning of natural kind terms is obviously in disagreement with the one given by Kripke and Putnam, which is widely accepted today. I have criticized the latter explanation in an earlier work (Ben-Yami, 2001).

Consider now sentences such as:

This animal is a tiger.
The tree over there is an elm.

That is, sentences in which natural kind terms are used as predicates. Since different people may identify the kind mentioned by different, logically independent properties, it is indeterminate which properties are attributed to the thing of which the natural kind term is predicated. Of course, in practice there is commonly a substantial overlap between the properties different speakers know particulars belonging to some kind have – we all know how a cow looks, say. But that is not always the case. What one knows of elms, genes or uranium changes radically from layman to expert. It is thus artificial to consider the use of natural kind terms as predicates a use in which a property, or some properties, are attributed to particulars. It is more natural to consider such a use *classificatory*, a use in which one classifies the thing referred to.

This is in contrast to the use of adjectives like 'dangerous' or 'tall' in 'This animal is dangerous' and 'That tree is remarkably tall'. Here it is most natural to say that one attributes properties to the mentioned animal and tree. It is also acceptable to consider the predicative use of common nouns that are not natural kinds terms a use in which a property is attributed to the particulars referred to. 'Student', say, in 'Jane is now a student' attributes to Jane the property of studying at a university or college.

The use of natural kind terms as predicates resembles in this respect the parallel use of proper names in the grammatically predicate position. When one says, 'The man sitting on the sofa is John Smith', one does not attribute any property to the man referred to, but identifies him. One says *who* that man is, while one says *what* a thing is when using a natural kind term as a predicate. The classificatory use of natural kind terms parallels the identificatory use of proper names. Natural kind terms are similar to proper names in this respect because both allow different people to identify the thing they refer to by logically independent properties.

Of course, one could argue that the sentence 'This animal is a tiger' attributes the property of *being a tiger* to the animal referred to. But I think that if this is maintained, one actually forsakes what we ordinarily understand by 'property', and instead identifies as a property whatever is predicated of a particular;[6] I prefer to preserve our ordinary use of 'property', vague as it may be. However, in case that definition of 'property' were accepted, we could also maintain that when one says, 'The man over there is John Smith', one attributes to the man mentioned the property of *being John Smith*. The parallel between natural kind terms and proper

---

[6] Kripke expressed a similar opinion about natural kind terms. In *Naming and Necessity* (p. 128) he wrote that 'in a significant sense, such general names as "cow" and "tiger" do not [express properties], unless *being a cow* counts trivially as a property. Certainly 'cow' and 'tiger' are *not* short for the conjunction of properties a dictionary would take to define them…'.

names would still be preserved; and natural kind terms would still be, when used as predicates, predicates of a special sort, with affinities to proper names.

What I intended to show in this discussion was that the semantic characteristics of some terms – natural kind terms – characteristics due to their function as referring concepts, make their use as predicates semantically distinguishable from the use of predicates whose function is to attribute properties. In this, natural kind terms resemble proper names, whose relevant parallel semantic characteristics are also explained by their referential function. This supports my claim that some concepts occasionally function as referring expressions, and that it is wrong to classify all general concepts as predicates, as the predicate calculus does.

The distinction drawn here, between the classificatory use of natural kind terms and the attributive use of other nouns and of adjectives, was foreshadowed by Aristotle. In the second chapter of his *Categories* he distinguished between what is predicable of a subject and what is present in the subject. The former class is constituted by common nouns (substance in the secondary sense, in his terminology), his examples there being 'man', 'animal', 'horse', 'tree' and 'plant' (*Categories*, chap. 5). The latter class is meant to include properties (Aristotle's terminology, 'being present in a subject', is a remnant of Platonic metaphysics), yet it also includes categories such as place ('in the market place') and time ('last year') (*Categories*, chap. 4). Although Aristotle did not distinguish in this case metaphysics from semantics, what can be retained from his distinction between being predicable of a subject and being present in a subject is, I think, the distinction drawn here between classification and attribution of properties.

3.5   Empty Names

The fact that in natural language common nouns are frequently used not as predicates but as referring expressions, as proper names are used, can account for the similar semantic characteristics of sentences with empty common nouns and sentences with empty proper names. (A referring expression is empty if it does not refer to any particular.) On the other hand, these characteristics and this similarity are problematic if common nouns are always predicative, as Frege has maintained.

Some time before the development of general relativity, astronomers believed that the perturbations in the orbit of Mercury are caused by an as yet unobserved planet. They dubbed that planet *Vulcan*. Later it turned out, of course, that there is no such planet. 'Vulcan' turned out to be an empty name, designating nothing.

Suppose an astronomer who has not heard about Einstein's discoveries claims that Vulcan is much heavier than Mercury. I think another astronomer, more updated than the former, would not be inclined to react by saying that what the first one said was mistaken or that he was wrong. In fact, we feel uncomfortable with the very question, 'Was what he said true or not?'. Even if it is correct to say that he was wrong, that is not the preferred reaction. The natural reaction in such a case would rather be to say that Vulcan does not exist, or that there is no such planet as Vulcan. We terminate the discourse about Vulcan, instead of continuing to use the name as if it designated some planet.

Similar remarks apply in the case of empty common nouns. Suppose some astronomers mistakenly believe that there are *several* planets between Mercury and the sun, and that while discussing their properties one of them claims that some of the planets between Mercury and the sun are heavier than the Earth. It would again be unnatural to insist that he was wrong; we should rather respond by saying that there are no such planets. Since the referring expression he used, 'the planets between Mercury and the sun', is empty, that should be made clear. That is the next move in the language-game, and not a move relating to the truth or falsity of what was said.

For a statement to be considered during a discussion as either true or false, we should succeed in doing with each expression contained in it what the expression is intended to do. If an expression is used to refer to particulars, then, in case of reference failure, the expression does not perform its function. In that case one should not continue the discussion by claiming that the statement has this or that truth-value. Rather, one should make clear that the statement involved reference failure. Common nouns in sentences of the form 'q A's are such and such' are used to refer to particulars. Accordingly, these sentences presuppose successful reference to A's. So, if 'A' does not designate anything, the reaction to an utterance of such a sentence should be similar to the reaction to a subject-predicate statement with an empty proper name as subject, a statement in which the proper name has a referential role.

By contrast, the sentence

(Some x)(Planet-between-Mercury-and-the-sun x & Heavier-than-the-earth x),

which is allegedly the correct translation of 'Some of the planets between Mercury and the sun are heavier than the earth' into the predicate calculus, is simply false, since it is false for every value of x. The problem of empty reference does not arise for this translation, since 'Planet-between-Mercury-and-the-sun' is a predicate, and not a referring expression. The alleged translation is closer in its meaning to the sentence 'There are planets between Mercury and the sun which are heavier than the earth', a sentence that does not presuppose reference to such planets, and which is simply false. This discrepancy between the alleged translation and the translated original indicates again the distortion involved in treating common nouns in quantified noun phrases as predicates.

This distortion will be made clearer if we compare the following two sentences and their purported translations:

1   Some witches are nurses
2   Some nurses are witches

Since conjunction is symmetric, they are both translated by

3   (Some x)(Witch x & Nurse x)

Now suppose both (1) and (2) are said of the real world (not of any fictional story). I think our natural response to (1) would be to say that there are no witches, while that to (2) would be that it is false. This is easily explained on my analysis of the function of common nouns in quantified noun phrases. In (1) the referring expression, 'witches', does not refer, whereas in (2) the referring expression, 'nurses', does, while 'witches' is used as a predicate. Since no nurse is a witch – there are no witches – the second sentence is false.

By contrast, (3) cannot explain the difference between our attitudes to the truth-values of (1) and (2), since the difference between reference and predication is lost in translation.[7] We again see that the analysis of common nouns in quantified noun phrases as referring expressions can explain linguistic phenomena that create difficulties if they are analyzed, following Frege, as predicates.

Strawson, in his discussion of presupposition in chapter 6 section 7 of his *Introduction to Logical Theory*, claimed that sentences of the form 'All A's are (are not) B's' and 'Some A's are (are not) B's' presuppose that there are A's. This claim, he maintained, is both intuitive and does justice to Aristotelian logic. Moreover, in chapter 6 section 8 of that book Strawson claimed that a sentence of the form 'All A's are B's' presupposes the existence of A's because the subject of that sentence, 'all A's', plays the referring role. Although I maintain that the common noun 'A', and not the grammatical subject 'all A's', is the referential expression in sentences of these forms, my position is clearly close to Strawson's. (I of course extend it, as Strawson would presumably do too, to other sentences of the form 'q A's are B's', where q is any quantifier: 'most', 'many', 'seven', etc.)

Strawson's position should be modified, however, in another respect. The presupposition of 'All A's are (are not) B's' and 'Some A's are (are not) B's' is not that A's *exist*, but that the speaker has succeeded to do with the expression what he or she attempted, i.e., *refer* to A's.[8] Now ordinarily, one intends to refer to existing

---

[7] The situation is similar if one uses generalized quantifiers. For instance, if we use binary quantifiers, then the translation of (1) is '(Some x)(Witch x, Nurse x)'. One would need to stipulate that in case the first predicate does not apply to anything in the domain of discourse the sentence has no truth-value, while this is not necessarily the case if the second predicate does not apply to anything in the domain. But this stipulation seems *ad hoc*, since the *two* concepts are used as predicates. On the other hand, as I have just shown, if only the first concept is a referring expression, as I claim, then the asymmetry of our responses is easily explained.

[8] I distinguish between names of fictional characters and names that do not denote anything. Names of fictional characters can very naturally be said to denote fictional characters. 'Hamlet', in 'Hamlet killed Polonius' and in 'Peter is as irresolute as Hamlet' is used to refer to Hamlet.

Referring is not a relation like touching: you can refer to a nonexistent thing, although you cannot touch it. Only a picture of relations dominated by relations like touching made philosophers assume that if two things are related, then both must exist. Referring is one among many counter-examples. One can *love* a fictional character or be *stronger* than a fictional character, a house can be *larger* than a fictional house, and so on.

This topic obviously requires more discussion, but this is not the place for it. I would just emphasize that it is not a *discovery* but a *decision*, albeit a natural one, to consider the mentioned use of fictional names referential (see Chapter 5). This use is similar in many respects to paradigmatic referential uses, although different from them in some significant

things, and therefore the presupposition of successful reference entails Strawson's presupposition of existence. But this is not always the case. One may speak of mythological characters, e.g., the Greek gods, and say that many of them were lascivious. Here there is no presupposition of existence, only of successful reference – as is also evidently the case when one says that some of Homer's heroes did not really exist.[9]

---

features. For instance, the use of a name to refer to a fictional character frequently cannot, because of its very nature, be substituted by pointing to the character referred to (unless the character is, say, a character in a film or a painting). The degree of similarity in use to paradigmatically referential uses should guide our decision.

[9] In section 11 of that chapter Strawson uses his claim, that existence is presupposed in a subject-predicate statement, to give 'a new edge to the familiar philosophical observation that "exists" is not a predicate': in a statement like (1) 'Many Homeric heroes existed' there is no presupposition of existence, and in this respect it is different, according to Strawson, from a subject-predicate statement such as (2) 'Many horses are brown'. Hence, 'exists' is not a predicate like 'brown'. If I am right, however, and statement (2) presupposes only successful *reference* to horses, not their existence, then in this respect it is not different from (1), which presupposes successful reference to Homeric heroes. Accordingly, 'exists' can still be considered a predicate, which seems quite plausible in this context. This is in accord with Strawson's own later views on existence – see his 'Is Existence Never a Predicate?'. (An advantage of my approach over his later view is that it avoids his *ad hoc* adaptations of what is exactly presupposed in such cases; see there.)

Chapter 4

The Sources of the Analysis
of Referring Nouns as Predicates

4.1 Frege

Given the plausibility of analyzing common nouns in quantified noun phrases as referring expressions, and all the distinctions mentioned in the previous chapter between predication and reference by means of concepts, their analysis as logically predicative obviously stands in need of justification. We should see whether we can find any good reasons for this *prima facie* implausible analysis. However, one can hardly find any attempt at such a justification in contemporary works in logic. After the predicate calculus had established itself in logic, its translation of common nouns by predicates has generally been simply taken for granted. Yet when Frege first analyzed common nouns in the subject position as predicative, this novel analysis could hardly have been so taken. We should now examine his reasons for this analysis.

Firstly, this analysis may have its roots in Aristotelian logic. Aristotle did distinguish between common nouns and adjectives, the former comprising a separate category (*Categories* 4; he conflates, however, semantics and metaphysics in his classifications). Nevertheless, common nouns, adjectives and verbs in the predicate position are uniformly treated in the Aristotelian syllogism. This encourages the view that common nouns in the subject position have the same function as common nouns in the predicate position, which have the same function as any general concept in that position. Until the end of the nineteenth century, this assimilation resulted more in the distortion of the semantics of predicates than in that of referring concepts. Both referring concepts and predicates were often considered names or denotative expressions (see, e.g., Mill's *System of Logic*, Book I, Chap. II, § 2). While this is true of common nouns in the subject position, predicates usually do not name anything. In 'Every philosopher is wise', 'wise' is not used to name or denote anybody, but to describe philosophers.

Frege inherited this conflation of reference and predication, but since his analysis of concept-words is more faithful to the semantics of predication than to that of reference, the understanding of reference suffered more this time.

Secondly, Frege, being influenced by the mathematical concept of a function, conflated the denotation/predication distinction with the singular-referring-expression/general-concept one.

In mathematical notation, the arguments of a function are always singular terms, and one naturally thinks of *singular* terms as denoting mathematical entities. For instance, in '$f(2)=2^2$', '2' is the argument, and it denotes the number 2; and in '$f(x)=x^2$', the argument 'x' is a variable, denoting nothing, but it can be replaced by any singular term that denotes a number.

Frege viewed logic through mathematical spectacles. As the subtitle of his *Begriffsschrift* declares, his formula language was 'modeled upon the formula language of Arithmetic'. He tried to interpret the concepts of natural language as embodying, perhaps in a misleading form, mathematical conceptions. His way of achieving that was by identifying concepts with functions; 'it seems suitable', he wrote, 'to say that a *concept* is a function whose value is always a truth-value' (1893, Book I, § 3; cf. his 1891, p. 15). Frege consequently replaced the subject/predicate distinction with the argument/function one (1879, section 9, and in many later writings). On his abandonment of the subject/predicate distinction he writes: 'In this I strictly follow the example of the formula language of mathematics, in which also to distinguish subject and predicate can only be strained' (ibid., § 3). This replacement brought with it the ascription of the denotative role to singular terms alone, as it is in mathematical notation. General concept-words were accordingly considered logically predicative (1892, p. 193; cf. 1976, p. 103).[1] As a result, since common nouns, being applicable to many particulars, are general concept-words, they were taken by Frege to be logical predicates.

The analysis of common nouns as predicative, even when they function as grammatical subjects, is already present in Frege's *Begriffsschrift* of 1879 (e.g., § 12). His arguments for this analysis, however, are found only in his later writings. In his *Die Grundlagen der Arithmetik* of 1884 (§ 47) we find the following argument, intended to show that in the sentence 'All whales are mammals' we do not speak about whales – i.e., 'whales' does not refer to whales – but about the *concept* of a whale:

> Indeed, the sentence (*Satz*) 'All whales are mammals' seems at first sight to be about animals and not about concepts; but if one asks, which animals then are talked about, even a single one cannot be indicated.

Of course, when we talk about whales, there is usually no whale around to which we could point. But all the same, the answer to the question, which animals are talked about?, is straightforward: whales. So what is the point of Frege's comment, that 'even a single one cannot be indicated'? Frege continues as follows:

> Suppose a whale were present, then nevertheless our sentence does not claim anything about it. One could not infer from our sentence that the animal present is a mammal,

---

[1] This claim is formally mistaken, since Frege thought of concept-words as denoting concepts, and of statements as denoting truth-values. This, however, is a technical expansion of our concept of denotation (*bedeutung*), while my use of the concept in the text in not technical.

without adding the sentence that it is a whale – something our sentence does not contain at all [*wovon unser Satz nichts enthält*].

The intended conclusion, that the sentence is not about whales, does not follow from this argument. If one said 'Peter is ill' and Peter were present, one could not infer that the man present were ill without adding that he is Peter – something our sentence does not contain at all. Yet 'Peter is ill' is about Peter, even according to Frege. Analogously, 'All whales are mammals' can still be about whales, even if one does not know of a certain animal that it is a whale. It seems that Frege confuses epistemology and semantics in this case. A sentence can be about certain particulars without somebody being able to say of a given particular in specific circumstances that it is among those the sentence is about – and this is the case with proper names as well as with common nouns.

Frege concludes his discussion thus:

> Generally it is impossible to speak of an object without designating or naming it in some way. But the word 'whale' names no particular.

This obviously begs the question. On the one hand, we can say that 'whale' designates all particular whales. On the other, we can say that although it does not specifically name any particular whale, it is used to speak about all whales. Frege fails to supply us here with a good justification of his position.[2]

In 'Über Begriff und Gegenstand' of 1892 Frege supplies further arguments for taking common nouns to be predicates, and not referring expressions, even when in the grammatical subject position (pp. 197-8). Firstly, he writes that even in a sentence like 'All mammals have red blood' the predicative nature of 'mammals' cannot be mistaken, since that sentence can be paraphrased as 'Whatever is a mammal has red blood' or 'If anything is a mammal, then it has red blood'. But one can use paraphrases both ways: why not say that the first sentence, where 'mammal' functions as subject, shows that in the paraphrases the concept 'mammal' is *not* used as a logical predicate? An additional reason is needed in order to justify taking the paraphrases as revealing the nature of the concepts in the paraphrased sentence, and not the other way around. Moreover, as was argued above (§ 2.3), two sentences can say the same thing although using concepts in different ways.

Secondly, Frege writes that if in

All mammals are land-dwellers

the phrase 'all mammals' were the logical subject, then its negation should be 'All mammals are not land-dwellers', which it is not. Accordingly, 'all mammals' is not the logical subject of that sentence. Frege repeated this argument in a note on page

---

[2] Frege repeated this argument, referring back to section 47 of the *Grundlagen*, in section IV of his review of Husserl's *Philosophie der Arithmetik* (page 83 in the English translation), and also on page 454 of his 'Kritische Beleuchtung einiger Punkte in E. Scröders *Vorlesungen über die Algebra der Logik*'.

441 of his 'Kritische Beleuchtung einiger Punkte in E. Schröders *Vorlesungen über die Algebra der Logik*'. He there used as an example the sentence 'All bodies are heavy', and argued that from the fact that the negation of this sentence is 'Not all bodies are heavy' it follows that neither 'all bodies' nor 'bodies' are the subject of that sentence.

But why should the negation of 'All mammals are land-dwellers' be of the form Frege specifies, in case 'all mammals' or 'mammals' are its logical subject? The negation of 'Peter and Mary are painters' is not 'Peter *and* Mary are not painters' but 'Peter or Mary is not a painter' or 'Peter and Mary are not both painters'. Yet, as I have argued in section 2.2, 'Peter and Mary' is surely the logical subject of that sentence: 'Peter and Mary' is used to specify two people of which something is said in the sentence. All that these examples show is that the syntax of a negation of a sentence with a plural subject is not that of the negation of a sentence with a singular subject.

Frege's reasons and arguments for not taking common nouns in the grammatical subject position to be referring expressions are therefore far from sufficient. It seems that Frege first formed, under the influence of mathematics that was discussed above, the artificial language of his *Begriffsschrift*, which mistakes such reference for predication. Only later did he try to justify this position, to which he was already committed by his calculus.

4.2   Russell and Bradley

Frege's reasons, however, were not the only ones responsible for logicians analyzing common nouns in quantified noun phrases as predicates, an analysis built into the predicate calculus. Russell's work had a more significant role in the almost universal acceptance of the calculus, and with it of this mistaken analysis. And Russell seems to have been convinced that such common nouns are logical predicates primarily by Bradley. While writing 'On Denoting', when he was already familiar with Frege's work, he considers the proposition 'All men are mortal' (p. 481). 'This proposition', he writes, 'is really hypothetical and states that *if* anything is a man, it is mortal. That is, it states that if x is a man, x is mortal, whatever x may be.' And Russell notes that this 'has been ably argued in Mr. Bradley's *Logic*, Book I, Chap. II.'[3] Russell consequently substitutes 'x is human' for 'x is a man', i.e., an adjective for a common noun, a substitution emphasizing that predication, and not reference, is involved. Thus, Russell's acceptance of Frege's predicate calculus' analysis of such common nouns as predicates was apparently made possible also by Bradley's arguments for their predicative nature. We should therefore consider Bradley's arguments.

---

[3] De Morgan (1847, p. 109) already argued that 'a conditional proposition is only a grammatical variation on the ordinary one ... Of the two forms, categorical ['Every X is Y'] and conditional ['If X, then it is Y'], either may always be reduced to the other'.

After arguing, on the basis of various psychological and metaphysical considerations, that all ideas are adjectives and all judgment hypothetical, Bradley maintains that 'more ordinary considerations might have led us to anticipate this result' (1928, p. 46). He then examines, among other sentences, universal ones. This is what he writes about the sentence 'Animals are mortal', or 'All animals are mortal' (p. 47):

> "Animals" seems perhaps to answer to a fact, since all animals who exist are real. But, in "Animals are mortal," is it only the animals now existing that we speak of? Do we not mean to say that the animal born hereafter will certainly die? The complete collection of real things is of course the same fact as the real things themselves, but a difficulty arises as to future individuals. And, apart from that, we scarcely in general have in our mind a complete collection. We *mean*, "Whatever is an animal will die," but that is the same as *If* anything is an animal *then* it is mortal. The assertion really is about mere hypothesis; it is not about fact.

And Bradley concludes (p. 48):

> Universal judgments were really hypothetical, because they stated, not individual substantives, but connections of adjectives.

We can take Bradley's 'answer to a fact' as meaning *designating* or *referring*. The first of his arguments against considering 'animals' as referring to animals is that since future animals do not yet exist, it is impossible to refer to them. His second argument is that since we do not have in our mind a complete collection, we cannot refer to a complete collection.

Let us consider Bradley's second argument first. In 'having in one's mind a complete collection' Bradley seems to mean that one thinks separately of each particular of the complete collection, perhaps on the basis of acquaintance or description. This would be the case when one refers, say, to all one's children. But why should this be a precondition for reference to a complete collection? We could on the contrary maintain that one can have a complete collection in mind in virtue of a concept which refers to all and only the particulars of these collection. I think this argument begs the question.

Let us next turn to Bradley's first argument: can we refer to non-existent individuals, to future ones for instance? Indeed, if the possibility of reference is to be admitted in such cases, it would on some occasions be a less paradigmatic case of reference than that to a present person (see Chapter 5). But first, even if this kind of reference is not admitted, that does not entail that in cases where this problem does not arise, as with common uses of the expression 'my children', the expression is not a referring one either.

Secondly, adopting the suggested adjectival form as perspicuously representing the actual logical relations in 'All animals are mortal' would not help us out of this conundrum. On Russell's version, we state that if x is an animal, x is mortal, whatever x may be. But then we use the variable 'x' or the quantified construction to refer to *future* individuals as well, in some sense of reference. So if a variable, or a quantified construction that uses variables, is allowed this kind of reference, it should be allowed to nouns as well, and the adjectival substitution is unjustified.

Lastly, there are good reasons for allowing the possibility of reference to future individuals. Consider the sentences

> All children born last year got the flu.
> All these children got the flu.
> All children that will be born next year will get the flu.

'These children', as frequently used in the second sentence, is a paradigmatic case of reference. Moreover, both the first and the last sentence are similar to it not only in their grammar, but in their method of verification as well: in all cases one should check, in some manner, all children belonging to a certain group, and see if they got the flu. For instance, one either asks all children born last year or their parents (1st sentence), or asks all the children pointed at or their parents (2nd sentence), or waits for at least a year and then proceeds similarly (3rd sentence). In addition, the situations described by all these sentences are very similar: each of a group of children having, at some time or other, the flu. Such considerations support the classification of all nouns coming after the quantifier in these three sentences as referential expressions. Accordingly, *pace* Bradley, it is legitimate to allow reference to future individuals.

To corroborate the argument of the previous paragraph, I proceed in the next chapter to a general discussion of reference.

Chapter 5

Reference

In this chapter I discuss our concept of reference. Although I shall try to clarify what reference consists in, this question will interest us only to the extent in which it helps either to allay qualms about some kinds of reference or demonstrate the parallel between singular and plural reference. My clarification of the nature of reference will therefore be only partial. I shall refer in two places to authors who discuss in more detail some central elements of the view I hold.

What does it mean, then, to say that a word is used to refer to certain particulars? Although reference is a key concept in semantics today, it is rarely explained. Its use, however, is obviously technical to some extent, so an explanation is called for.

The explanations one does find in contemporary literature on the philosophy of language are often such as the following one. A term used in a sentence refers to an object, it is maintained, if it contributes the object to the content of the sentence. This is no good. If we understand what is meant by 'to contribute an object', it is because we take 'contribute an object' as synonymous with 'refers to an object'. Even if this formulation is a correct definition, it does not help to clarify the concept of reference. It is merely a semblance of an explanation.

Other popular explanations are roughly of the following form: a term used in a sentence refers to an object if it makes the object relevant to the truth-value of the sentence. Ignoring some surmountable difficulties (e.g., reference in questions or commands, which do not have a truth-value), the main problem with this explanation is that it does not specify the intended sense of 'relevant'. If one says, 'Paul is out', then Jane might be relevant to the truth-value of this statement, since Paul is often with her. And if one says, 'Peter is our dean', then Jim is relevant to the truth-value of that statement, since if he had been the dean it would have been false. But neither Jane nor Jim were referred to in these examples. Particulars relevant, *in some sense*, to the truth-value of what is said, in either actual or possible circumstances, are not necessarily referred to in the statement. So what is the intended sense of 'relevant' in the above explanation? – It is, obviously, the one which makes only the object *referred to* relevant, and again we are left with no explanation.

A well-known attempt to explain reference, which does not consist in an obscure tautology, is Quine's. Quine attempted to explain meaning and reference by means of Skinner's concept of a conditioned reflex (Quine, 1960, pp. 80-82). The meaning of sentences and words is ultimately derived, according to Quine, from the meaning of what he called *observation sentences* (ibid., § 10), whose meaning is ultimately their *stimulus meaning* (ibid., § 8). Stimulus meaning has two parts, affirmative and negative. The affirmative stimulus meaning of 'Rabbit', say, consists in the class of all stimuli that would prompt the speaker's assent, while its negative stimulus meaning consists in the class of those that would prompt the speaker's dissent (cf. ibid., where the definition is elaborated a little). If we ignore Quine's commitment to stimuli – his materialist version of sense-data – we can say that according to him 'rabbit' refers to rabbits because we are prompted or disposed to assent to its assertion when we see a rabbit, and to dissent from it when we do not see any. (I also ignore Quine's alleged indeterminacy of reference, irrelevant to this discussion.) Quine has continued to hold this conception of meaning and reference ever since *Word and Object* (cf. his 1995, passim.).

The fundamental mistake in this conception of meaning and reference is that its presupposed dispositions do not exist. In most circumstances in life, if someone were to approach us and point, say, to our shirt and say 'shirt', we would not be prompted or disposed to assent in any way; and even less so would we be disposed to dissent were he to point to our shirt and say 'rabbit'. The dispositions Quine presupposes exist to some extent only when we attempt to *teach* our language; but teaching presupposes a competent use of language, a use in which words have a reference which teaching tries to match. Accordingly, reference is grounded in a use from which Quine's dispositions are missing. His conception of meaning and reference is therefore misguided. Quine almost reduces his unrealistic conception of language to absurdity when he claims that 'the ideal experimental situation' for studying an unfamiliar language is 'one in which the desired ocular exposure concerned is preceded and followed by a blindfold' (1960, p. 32): in these circumstances, the use of language will have next to nothing to do with its actual use.

Quine was correct, however, in his attempt to explain reference by focusing on the use of words. His mistake was in what he took that use to be grounded on. If we wish to explain what reference consists in, we should consider a more realistic case of word use.

Such a use is described by Wittgenstein, in the second section of his *Philosophical Investigations*:

> Let us imagine a language for which the description given by Augustine is right. The language is meant to serve for communication between a builder A and an assistant B. A is building with building-stones: there are blocks, pillars, slabs and beams. B has to pass the stones, and that in the order in which A needs them. For this purpose they use a language consisting of the words "block", "pillar", "slab", "beam". A calls them out; B brings the stone which he has learnt to bring at such-and-such a call.

Unlike Wittgenstein, we do not need to assume that the language he describes is a complete primitive language; it is enough that this is a use to which our language can be put.

As Wittgenstein writes, this is a use for which the description given by Augustine (*Confessions*, I. 8.) is right. This use inspires 'a particular picture of the essence of human language. It is this: the individual words in language name objects' (Wittgenstein, ibid., § 1). It is, we can say, a referential use of words. And Wittgenstein writes later on (§ 37):

> What is the relation between name and thing named?—Well, what *is* it? Look at language-game (2) or at another one! there you can see the sort of thing this relation consists in.

A says 'slab'; B brings him a slab; A takes it and continues with his work. A says 'slab'; B brings him a block; A doesn't take it, he stops his work and says to B, 'No, slab!'; B apologizes and quickly fetches a slab. The fact that utterances of 'slab' are connected with this sort of behavior is what it means that the word is used to refer to slabs.

Similarly with proper names. 'Please give this book to Peter': if you give the book to someone else, corrective behavior will ensue. You inform me: 'I gave Peter the book'; I later call *Peter* and ask him how he found the book I had earlier asked him to read. 'Where is Peter?' – 'There he is', and I point to *Peter*. Such connections between words and behavior constitute reference.

These examples require some methodological clarifications. Firstly, they are examples, and not a general description of what reference consists in. By contrast, all explanations mentioned above, which I rejected, attempted to give a general description. Such a description, if possible, is surely preferable to examples. However, I think explanation by means of examples is the form explanations should, by and large, take when basic concepts such as reference or naming are concerned. No general concept that can be used here to give a general definition will be clearer than that of reference. Compare the similar approach of Grice and Strawson in their 'In Defense of a Dogma', pp. 147-52, when criticizing Quine's implicit demand of a satisfactory explanation of the concept of analyticity, that it should take the form of a strict general definition.

Secondly, it might be remarked that the kind of behavior I describe does not necessarily accompany the referential use of words. For instance, even if B brought A a block when the latter said 'slab', A might think that he can make some use of a block too, and continue with his work without correcting B. All the same, A referred to a slab, and not to a block. Thus, one might conclude, the kind of behavior which *usually* accompanies the referential use of words does not accompany it *necessarily*, and it cannot therefore be what reference consists in.

But this argument tacitly presupposes the invalid inference from the true statement, that exceptions are always possible, to the false one, that it could be that all cases be such exceptions. The meaningful use of language presupposes an actual form of life, which endows language with meaning.

The justification of this last claim would obviously demand an extensive discussion. Instead of supplying it I refer the reader to Wittgenstein's *Philosophical Investigations*.

Thirdly, a point more directly related to the subject of this work: these examples are paradigms of a referential use of words, but it is unclear how they should be extended to other cases, which may resemble them to different degrees. Thus, what reference consists in is not precisely determined by these examples. That, however, is how it should be: there is no unique way in which these paradigms should be extended. Our concept of reference is a vague family-resemblance concept. Any definition that draws exact boundaries would substitute a different concept for ours, a substitution which may be desirable for some purposes, but which is not the aim of our present inquiry.

In the rest of this chapter I shall be concerned with various examples of the use of words, uses which will move from the obviously referential to what can hardly be considered as such. My examples will resemble paradigmatic referential uses in various ways and to different degrees. The decision, occasionally compelling, to consider a use referential, will be based on its similarity to paradigmatic cases of reference, similarity both in grammar and use. This procedure will serve us in drawing the vague boundaries of our concept of reference.

One picture of reference is perhaps that of pointing by means of words; Wittgenstein's A could point to a slab instead of saying 'slab'. Reference by words is a linguistic extension of pointing. We have already seen this picture at work in a quotation from Max Black (p. 11). Now pointing, like reference, is a complex practice. It has various consequences, depending on the circumstances of its use. A child points at a toy and his father gives it to him. Jane points to a shelf while saying 'Put it there', and Peter puts the vase on that shelf. And so on: a multiplicity of uses, resembling each other in various ways. Moreover, a movement of a hand is pointing because of its place in what we do; in a different context, the same movement may be part of a dance, say. We should not assume that we reduce reference to something simple when we consider it pointing by means of words. All the same, this picture can serve us when we discuss cases of reference. Paradigmatic referential uses of words are those where we could, instead, point with our finger to the object referred to, if it were present.

One paradigm of referential use is the use of 'Mary' in the following piece of conversation:

'Who is leaving?' – 'Mary is.'

Here Mary was referred to by means of the word 'Mary'. Instead of saying 'Mary', one could point to Mary if she were present.

Now we could also use 'she', 'they', 'John and Mary', 'the children', etc. in a way similar to that in which 'Mary' was used in the example above. Accordingly, all of these should then be considered referential. In this kind of use there is no distinction relevant to reference between singular and plural expressions.

Should the use of 'Mary' in

John saw Mary in the supermarket

be considered referential? I think it is more naturally considered referential if the sentence is said in response to the question, 'Whom did John see in the supermarket?' than as a response to the question 'Who saw Mary in the supermarket?'. The first answer, but not the second, could be substituted by pointing to Mary. However, since the statement made describes the same event in both cases, it is legitimate to consider the use of 'Mary' in both as referential. All the same, it is important to realize that the concept of reference is thus somewhat extended. As regards plural versus singular expressions, both can be used the way 'Mary' was used in this example. John could have seen *Tom and Mary*, or he could have seen *the children*; and, if we point to Tom and Mary, we can say that John saw *them* in the supermarket.

Is the use of 'Mary' in the question

Have you seen Mary?

referential? I do not think we would ordinarily be inclined to consider it such, but it is of course legitimate to call it referential in some technical sense. There are various reasons for doing that. For instance, the use of 'Mary' in 'I *have* seen Mary' in an answer to that question, or of the pronoun 'her' in 'I *have* seen her', is referential, according to the standards of preceding paragraphs. So again, in some weaker, extended sense, this use can be considered referential as well. As long as we are not misled by this decision into thinking that we are guided by some essential feature that constitutes reference, we are free to decide either way. As for singular versus plural expressions, no distinction relevant to reference exists in these cases either.

Let us now consider the sentences:

The man standing by the window must be a philosopher.
The tallest man in the world must be miserable.

The use of 'the man standing by the window' in the first sentence is naturally considered referential. It can be accompanied by pointing, for instance. Now because of the syntactic and semantic similarities between the two sentences, we can consider the use of 'the tallest man in the world' in the second sentence referential as well, although the first sentence but not the second usually assumes some kind of acquaintance with the man referred to. For instance, speakers will usually be able to identify the man referred to in the first sentence by means independent of the description used, while they must rely on the description in the second sentence in order to identify the man referred to in it. In the second example we are further removed from the paradigmatic cases of reference we started with, and accordingly the decision to count as referential the use it makes of the definite description is less compelling. Indeed, some prefer to distinguish between referential and attributive uses of definite descriptions, and to consider the second use above attributive (Donnellan, 1966). I think, however, that it would be more natural to distinguish between reference accompanied by acquaintance and merely descriptive reference. But in any case, singular and plural expressions can both be used in this way, so no distinction with respect of reference is justified between them.

Consider next this short conversation:

'Who is this man?' – 'It's John Smith.'

I do not think there is sufficient justification for counting the use of 'John Smith' here as referential. Attention was not drawn to John Smith by means of his name, and to find out whether the answer was correct one should only inquire how the man indicated is called. In addition, one cannot see how the word 'John Smith' could have been replaced here by that paradigm of reference, pointing: to whom should one point? Pointing to the man indicated will not supply any answer, and pointing to another person would be a mistake. Thus, it is more plausible to maintain that 'John Smith' is used here to identify someone, and not to refer to anybody.

To support this claim, we can use a principle which Geach ascribed to Buridan (*Sophismata*, chap. vi, sophism v) and named *Buridan's Law* (Geach, 1962, p. *xi*; cf. Geach, 1961-62, pp. 94-5). The principle says that the reference of an expression should be specifiable in some way that does not involve first determining the truth-value of the proposition in which it occurs; determination of reference (which is part of the determination of meaning) is a precondition of determination of truth-value.[1]

Now in case the answer to the above question is wrong, i.e., the man indicated is not John Smith, it seems strange to maintain that John Smith was referred to; the speaker simply misidentified the man indicated. Thus, according to Buridan's Law, it is plausible to maintain that even if the answer is true, 'John Smith' is not used to refer to anybody.

Other expressions can be used in this way as well. For instance:

This is *the prime minister.*
These are *the people who will escort you to the airport.*

I think it is strained and unjustified to consider the use made of these italicized expressions referential. They are more naturally considered identificatory. However, as demonstrated, both singular and plural expressions are used in this way. So whether or not one decides to count these uses referential, that should not justify any distinction between singular and plural expressions. (See more on identity in section 11.5.)

Lastly, it would be odd to consider as referential the use of the expressions 'Mary!' or 'Boys!' in exclamations or to draw attention. But even here we see that singular and plural expressions have parallel uses.

We can move away from paradigmatic examples of reference in a route different from the one followed above. Consider these two sentences:

---

[1] Buridan actually formulated a somewhat different principle in that place: 'a proposition presupposes the signification of its terms. For it is first required that the terms be imposed to signify before any proposition is formed from them' (1966, p. 167). That is, Buridan maintained that the reference of an expression should be specifiable in some way that does not involve first determining the *meaning* of the proposition in which it occurs. However, Buridan uses his principle to reject a sophism in which the reference of a term is determined by the truth-value of the proposition in which the term occurs; i.e., he in fact uses the principle Geach formulated. Geach's ascription of his principle to Buridan is therefore acceptable, and I shall continue to refer to it as 'Buridan's Law'.

The man carving this statue is a fine artist.
The man who carved this statue was a fine artist.

The uses of the definite descriptions in these two sentences are very close. For instance, the truth-value of both can be determined in the same way: by examining the corresponding statues. Moreover, the implications of the two sentences on what we may do are also similar. We may look for other statues by the same artists, we may praise their work, we may try to imitate it, we may be interested in learning more about the artists, and so on. However, there might be one significant difference between the sentences: the second artist may no longer exist.

Should this difference influence our decision, whether to count the use of the second definite description referential? I do not think it should. The two sentences are grammatically very close to each other; it may be unknown to speakers and their audience whether the second artist is still alive, and it would be strange to let this unknown fact determine whether the definite description is used referentially; the understanding of both sentences is manifested in similar ways; they can be put to similar uses. If our picture of reference is that of pointing by means of words, which is often pointing *in absentia*, why cannot the absence be permanent? I find it most natural to extend the possibility of reference to people and things that no longer exist.

Should reference to future individuals be allowed as well? I discussed this question at the end of the previous chapter, and I have little to add in this place. The resemblance to reference to existing individuals is diminished, but it still exists: in grammar, method of verification, and the use that can be made of the sentences. So it is legitimate, although less compelling, to consider this use referential.

One thing, however, deserves notice: even if we do not count as referential the use, say, of the definite description in 'The children who will be born next year will be vaccinated', this does not entail that we have to consider it predicative. To consider the use of any concept as either referential or predicative would be to impose an unjustified dichotomy on language. In case the use of the definite description above is not considered referential, we would do best to describe its similarity to and difference from referential uses, and to introduce some technical term applicable to this and similar uses. But again, I think the similarity to paradigmatic cases of reference justifies countenancing reference to future individuals, as long as we are conscious of the distinctions between this use and other, more paradigmatically referential ones.

Reference to individuals that no longer exist resembles reference to fictional individuals, those who never existed. I have already made some notes on this topic (note 8, p. 39). Here I would add that frequently it is uncertain whether some characters are fictional or real – that is the case with many characters in the Scriptures, e.g., Abraham and Moses. Accordingly, if we allow reference to past individuals – which I claimed is a most natural decision – it would be awkward not to allow reference to fictional ones as well. Not allowing such reference would make the answer to the question, whether a use of a term is referential, sometimes depend on contingent results of future research, and not on facts about its actual place in our linguistic and non-linguistic behavior. Again – such a decision is legitimate; but it would be an awkward one, and far from compelling.

As regards plural versus singular reference, there is no relevant distinction between the use of plural and singular expressions when past, present or fictional individuals are concerned.

The last use I shall discuss in this chapter is that of nouns as in the following examples:

> Wisdom is a virtue.
> Philosophy has lost much of its glamour.

Should we consider the use of 'wisdom' and 'philosophy' in these examples referential? One may be inclined to say we should, since these nouns are used here to *mention* wisdom and philosophy, or because we want to *say something about* them. But this fact is, in this case, the same as the fact that 'wisdom' and 'philosophy' are the subjects of these sentences. We should rather inquire what does the fact that we speak about wisdom and philosophy in these cases consist in.

To say that wisdom is a virtue is roughly synonymous with saying that wise men are in this respect virtuous, or that it's good to be wise. To say that philosophy has lost much of its glamour is more or less like saying that people do not find philosophizing as glamorous as people once did. In order to understand what these sentences mean, we should know what it means to be wise or to philosophize. And 'to be wise' and 'to philosophize' are now used as predicates, to say something about individuals referred to be other means. That is, the meaning of 'wisdom' and 'philosophy' is determined by the predicative use of these concepts. Their predicative use is prior from an explanatory, and therefore also a semantic, point of view. In this sense 'wisdom' and 'philosophy' are *semantically derived nouns*.[2]

The use of semantically derived nouns is far removed from paradigmatic referential uses. Semantically derived nouns do not point to anything, but rather abbreviate a more elaborate use of predicates. The place of their use in our behavior, linguistic or other, has little in common with that of paradigmatic referential uses. To count this use referential, in the same sense that the use of 'Mary' in 'Mary has arrived' is, may easily lead to metaphysical fantasies. The only justification for considering it referential is syntactic, not semantic. In this work I do not count this use among those I consider referential.

However, although the use of semantically derived nouns is only faintly related to paradigmatic referential uses of common nouns from a *semantic* point of view, their incorporation in the same place in the *syntactic* framework is conditioned by the applicability of parallel syntactic transformations and derivations to them. Consequently, a deductive system that applies to *common nouns* that have a specific

---

[2] A *semantically* derived noun does not have to be *grammatically* derived as well (although that is usually the case) – 'courage' can be considered semantically derived from 'brave' as well as from 'courageous'.

place in syntax, would also apply to *semantically derived nouns* that have the same place.

I shall return to this point in section 8.6. But one further note on semantically derived nouns is appropriate here. I tried to explain the nature of reference by describing the place of referring expressions in our behavior, primarily our *non*-linguistic behavior. In this way, the expressions that are used referentially are identified primarily by *extra*-linguistic criteria. By contrast, some other writers try to determine which expressions are used referentially by *intra*-linguistic criteria. Dummett, for instance (1981, pp. 57*f*), and, following him, Hale (1987, § 2.I), try to characterize referring expressions by the kind of inferences in which they participate. As a result, terms like 'sliminess', 'shininess' and the like turn out to be names of objects on their criteria (Dummett tries to find a way to avoid this result (ibid., pp. 70-80), while Hale accepts it (ibid., pp. 32-41)). This result demonstrates that their approach is misguided. Referring is one of the ways in which language connects to *non*-linguistic reality, and it should be explained by that connection. As Wittgenstein wrote (*Philosophical Investigations*, § 11):

> Of course, what confuses us is the uniform appearance of words when we hear them spoken or meet them in script and print. For their *application* is not presented to us so clearly. Especially when we are doing philosophy!

The application of semantically derived nouns is determined via the non-referring expressions employed to paraphrase and explain their uses. That is why they are not referring expressions and name no object.

With this I conclude my discussion of reference. This discussion was not meant to be exhaustive. Other uses of names and other expressions are more or less closely related to those mentioned above, and some can therefore be considered referential – as long as their distinctions are also noted. But since my main aims were to compare singular and plural reference, and to allay some apprehensions caused by misconceptions about reference, I do not think I need to make my discussion of reference any more complete.

I thus end Part I of my book with the following conclusions. The notion of *plural* reference is straightforward. The parallels between referential uses of singular expressions and some uses of plural expressions supply us with a very good reason to consider the latter referential. Attempts to reduce apparent plural reference to semantic phenomena of other kinds have not been motivated by any linguistic phenomenon, and are at least implausible. There are good reasons for considering the use of common nouns in quantified noun phrases referential in many cases. This analysis is supported by several linguistic phenomena. Frege's arguments and reasons for the analysis of common nouns in these uses as predicative are far from compelling. With these conclusion we can now proceed to discuss the nature of quantification in natural language.

PART II
QUANTIFICATION

Chapter 6

Quantification: Natural Language versus the Predicate Calculus

6.1 The Nature of Quantification, and the Differences between Its Implementations

The absence of plural referring expressions from the predicate calculus forces quantification to function in the calculus in a way that is significantly different from the way it functions in natural language. When we quantify, we refer to a plurality of particulars, and say that specific quantities of them are such-and-such; quantification involves reference to a plurality. Natural language accomplishes this kind of reference by means of plural referring expressions, which designate the plurality, or pluralities, about which something is said. And by using different expressions, natural language can refer to different pluralities. By contrast, since the predicate calculus uses concepts only as predicates, it has no plural referring expressions. The plurality about which something is said by its sentences has to be presupposed, and different sentences cannot specify different pluralities. In natural language, pluralities are introduced and specified by means of plural referring expressions; in the predicate calculus, a plurality, which is unspecified by the sentence, is introduced by presupposing a domain of discourse.

In order to speak of pluralities natural language sentences presuppose no domain of discourse, in the technical sense in which this concept is used in predicate logic semantics. A domain of discourse is a necessary component of the semantics of the predicate calculus, which has no parallel in the semantics of natural language.

Of course, context is needed in order to determine, for instance, which students one refers to when one says, 'Some students were late'. But the context does not do that by first determining a domain of discourse, a domain which may also contain some particulars that are not students. Similarly, when one says, 'John was late', the context determines which John one refers to, without determining a domain which may also contain some unmentioned Pauls and Peters. In the predicate calculus, the context determines a domain of discourse, which may contain many particulars that will not be mentioned at all. In natural language, the context directly determines the reference of the concepts used.

This semantic difference results in a syntactic one as well. If the plurality is referred to by some plural referring expression, the quantifier has to be related in some syntactic way to the plural referring expression, in order to indicate the plurality of which a quantified claim is made. Consequently, in natural language the quantifier is attached to a noun that is used to refer to a plurality, and together they form a noun phrase. However, if no expression is used to refer to a plurality, but the plurality is presupposed by the quantified construction, then the quantifier does not have to be attached to any specific component of the quantified sentence. Consequently, in the predicate calculus the quantifier operates on a sentential function.

I shall now demonstrate the above by examining an example. In my example I use unary quantifiers. Indeed, at present binary and restricted quantifiers are often used in linguistics and logic, translations by their means being superior in various respects to those made with Frege's original unary quantifiers. But the essential differences between quantification in natural language and quantification in the predicate calculus are equally demonstrable by all kinds of quantifiers. And since unary quantifiers are more widely known, and quantification by their means is still considered the standard or canonical mode of quantification, I prefer to use them in my demonstration below. Binary and restricted quantifiers will be introduced and discussed later, in section 6.4.

Let us then proceed with an example:

1   All men are mortal

is usually translated by

2   (all x)(Man $x \rightarrow$ Mortal x).

This translation departs in several ways from the semantics of the translated sentence. Firstly, in sentence (1), 'men' is used to refer to all relevant men and to them alone, while 'mortal' is a predicate, used to attribute a property to men. In (2), on the other hand, both 'Man' and 'Mortal' are predicates – as we have already noted, the semantic distinction between reference with a concept and predication with one is lost.

As a result of these differences between (1) and (2), the following additional difference arises, concerning the way the quantifier functions in each. In (1) 'all' is joined to the referring expression 'men' (together they form the noun phrase 'all men'), and it determines that the predicate should apply to all the particulars that *the term 'men' designates*. By contrast, in (2), 'all' is joined to the variable 'x', and it determines that a complex predicate, the sentential function '(Man $x \rightarrow$ Mortal x)', should apply to all the particulars *in a presupposed domain*. Sentence (2) does not *specify* any plurality of particulars, but *presupposes* one. In both natural language and the predicate calculus, quantifiers determine to how many particulars from those referred to a predicate should apply. But while plural reference in the calculus is introduced by attaching a quantifier and a variable to a sentential function, in natural language it is made by general nouns, to which quantifiers therefore attach.

Variables resemble plural referring expressions in some respects. The plural reference involved in a quantified sentence of the predicate calculus is introduced by the quantified construction itself, and not by any specific expression in it. Variables are needed only in order to disambiguate quantified statements about relations between different particulars. For instance, sentence (2) could also have been written as '(all)(Man → Mortal)' without any ambiguity; but variables of some sort are already essential for a statement like

(every *x*)(some *y*)(Man *x* → Mother(*y*, *x*)).

Variables are needed in this case in order to determine the way the relation holds; i.e., whether we are saying that every man has a mother, or that every man is somebody's mother.

However, the variable stands in sentential functions in places which are occupied in non-quantified sentences by singular referring expressions. Moreover, the kind of particulars referred to in the domain is sometimes limited by changing the variables used. For instance, one sometimes uses x_1, x_2, x_3 etc. to denote substances, e_1, e_2, e_3 etc. to denote events (Davidson, 1980, essay 7), t_1, t_2, t_3 etc. to denote times, and so on. Accordingly, plural reference in the calculus *can* be seen as made by bound variables, which can then be considered plural referring expressions of a simple form. We thus find logicians and philosophers claiming that bound variables 'refer to entities generally' (Quine, 1948, p. 6) or that 'a variable refers to its values' (Quine, 1987, p. 180).

My observation in the first part of this work, that the predicate calculus lacks plural referring expressions, might have led some readers to think that if we added such expressions to the calculus it would then require only few if any additional modifications. However, the way quantification functions in the calculus shows that its semantics is fundamentally different from that of natural language. In natural language a plurality of particulars is introduced by means of a plural referring expression; while in the predicate calculus the plurality is introduced by means of a bound variable. In the predicate calculus, quantifiers specify how many particulars from a presupposed domain have a certain property; the quantifier in natural language, by contrast, specifies how many particulars of a plurality introduced by a general term have a certain property. Already for these reasons, if one wanted to develop an artificial language that could represent the semantics of natural language, one should depart from the predicate calculus to such an extent that the outcome could hardly be considered a modification of the latter.

The predicate calculus cannot even be seen as a simplified model of a fragment of natural language. In this it contrasts with the propositional calculus. The latter *is* such a simplified model, which depicts, for some sentential connectives, some of the ways they function in natural language. If the case with the predicate calculus were analogous, some quantifiers and nouns should function in some of their uses in natural language in the way their analogues function in the calculus. As I have argued, however, this is not the case.

6.2   Definite and Indefinite Noun Phrases

I shall now consider noun phrases in general, and contrast those containing a quantifier with those formed with other determiners. My claim will be that while some noun phrases are referring expressions, others, although containing a part which is used to refer, are not referring expressions themselves.

It should be noted once again, however, that expressions which are used in some statements to refer to certain particulars, can be used non-referentially in other statements, without any change of meaning. Contrast, for instance, the non-referential use of the definite description in 'Napoleon was the greatest French soldier' with its referential use in 'The greatest French soldier died in exile' (Strawson, 1950, p. 1). My discussion below is therefore limited to some uses of noun phrases, primarily those where they function syntactically as subjects of one-place or many-place predicates. (I use 'subject' to designate in sentences both the grammatical subject and the grammatical object, either direct or indirect. For instance, in 'She gave him the money' I call 'she', 'him' and 'the money' the subjects of the three-place predicate 'give'.)

Consider first a noun phrase consisting of (1) a common noun in the singular or plural; (2) optional grammatically attributive adjectives, relative clauses, etc. that modify it; and (3) a determiner like 'the', 'these', 'her' etc., that turn the whole noun phrase into a definite description. Examples are: 'the fat man/men', 'this child'/'these children', 'my sister/s' 'the tallest man in the world', 'the nice men who came to visit you yesterday'. *Such noun phrases, when used as subjects, are referring expressions.* To use them is to point by means of words. I shall call the determiners 'the', 'this', 'that', 'these', 'those', 'my', 'her', etc. *definite determiners.* I shall call a noun phrase consisting of a definite determiner followed by a general noun *a definite noun phrase.* (This classification is language-dependent; possessive adjectives, for instance, are not determiners in Hebrew.)

By contrast, a noun phrase formed from a *quantifier* followed by a noun is *not* a referring expression. The *noun* (which may be a common noun modified by attributive adjectives etc.) in that noun phrase is used to refer to the relevant particulars; the quantifier determines to how many of these particulars the predicate should apply. (This rule applies to sentences with a single quantified noun phrase, and will be elaborated below (§ 7.2) for those with several ones.) For instance, when I say 'Some students are late', 'students' refers to *all* students under discussion, and I say that some of them are late. In making that statement I am *not* referring to *some* students, but to *all* of them. Similarly, in 'Most of my new students are keen on this subject', 'my new students' is the referring expression, designating all my new students.

My claim that in 'Some students are late' the noun 'students', and not the noun phrase 'some students', is the referring expression, is in agreement with Aristotelian Logic's identification of logical subjects. The noun, and not the noun phrase, was considered the logical subject of such propositions – the '*S*' in '*SiP*', for instance. The quantifier was considered a logical constant, a syncategorematic term, signified in the formalization (together with the mode of predication) by the copula, '*i*' in this case.

To support my claim that in such sentences the noun, and not the noun phrase, is the referring expression, consider again Buridan's Law (p. 52; cf. its application by Geach in his 1962, § 4). Whether a word or an expression is used to refer and what they refer to should be determined without reliance on the truth-value of the statement made; determination of reference is a precondition of determination of truth-value. Accordingly, the reference of the referring term in 'Some students are late' should be the same, whether the statement made is true or false. Suppose the statement is false. Which students should 'some students' denote then, if it is a referring expression? There is no set of students which seems a plausible candidate. Accordingly, even when the statement is true, the phrase 'some students' does not denote any set of students, in particular it does not denote those who are late. (Although speakers, assuming the statement to be true, may continue to talk about those students who were late, referring to them by 'they', say). By contrast, if we take '*students*' to be the referring expression, which denotes *all* relevant students, no such difficulty arises.

This argument applies in this form to all quantifiers apart from 'all'. If we want it to apply to 'all' as well, we can consider sentences like 'Most or all students were late'. Moreover, we have no reason to assume that the function of nouns or of quantifiers changes when 'all' is substituted for any other quantifier in a sentence.

Aristotle seems also to have been of the opinion that the noun (the subject, in his terminology), and not the noun phrase, is the referring expression in universal propositions. For he writes (*On Interpretation* 7, 17b12) that 'the word "every" does not make the subject a universal, but rather gives the proposition a universal character.'

I shall call the quantifiers 'all', 'some', 'most', 'seven', 'more than eleven', 'infinitely many', etc. *indefinite determiners*. A noun phrase consisting of an indefinite determiner followed by a general noun is *not* a referring expression. The general noun in it is the phrase used to refer. I shall call such a noun phrase *an indefinite noun phrase*.

There is thus a considerable semantic difference between definite and indefinite noun phrases. In some languages (e.g., Hebrew) definite and indefinite determiners have also a markedly different syntax.

6.3   Geach and Strawson on Plural Reference and Quantification

Peter Geach (1962, pp. 180, 188) ascribed to Aquinas an analysis of the semantics of quantifiers on the same lines as mine. This ascription is based on the following line from the *Summa Theologiae* (Ia., Question 31, Article 3, p. 92):

Dictio vero syncategorematica dictiur quae importat ordinem praedicati ad subjectum sicut haec dictio 'omnis' vel 'nullus'.

Geach understands Aquinas as saying that determiners (which Geach calls 'applicatives') – like 'all' and 'none' in this quotation – show 'how the predicate goes with the subject' (Geach, 1962, p. 188). Although this can be interpreted as containing an analysis of quantifiers similar to the one I suggest in this book, I find such an interpretation insufficiently supported by the textual evidence.

Geach himself, however, adopts the view he ascribes to Aquinas. He considers (ibid., § 105) the sentence '$F(q A)$', where 'F' is a predicate, 'q' a quantifier and 'A' a substantival general term (I adapt Geach's terminology to mine). In such a sentence, Geach maintains, 'A' has the role of a name, of a logical subject standing for individual things. This view, he emphasizes, is in disagreement with Frege's, who maintained that 'A' in such a sentence stands for a concept. I am in agreement with Geach on these points.

However, Geach thinks (ibid.) that this conception

> guides us to a full acceptance of Frege's view about referring phrases ... We should read 'F(every A)' and 'F(some A)' as got by attaching the different predicates 'F(every _)' and 'F(some _)' to 'A', not by attaching the predicable '$F($)' to two different quasi subjects 'every A' and 'some A', which refer to the things called 'A' in two different ways.

I agree, of course, with Geach that the second alternative should be rejected. But I do not find it semantically rewarding to 'read "F(every A)" and "F(some A)" as got by attaching the different predicates "F(every _)" and "F(some _)" to "A"'. Geach reaches this position as a result of his decision, unjustified from a logical point of view, to analyze every proposition as a predicate attached to a subject (ibid., §§ 18, 26). I think we shall have a more adequate conception of the semantics of '$F(q A)$' if we see it as composed out of three elements with different semantic roles: 'A' refers to particulars, 'F' is predicated of particulars, and 'q' determines to how many of the particulars denoted by 'A' the predicate should apply. (Although this analysis is also insufficient for my purposes; cf. my analysis below, section 8.6.)

Geach's view of the semantics of common nouns and quantification being close to mine in important respects, I shall note here some differences between us on this subject.

Firstly, by contrast to Geach, who does not ascribe any semantic role to the copula, I think copulas – affirmative, negative, and other – have an essential semantic role in determining the mode of predication (see sections 7.4, 8.6).

Secondly, unlike me (see Chapter 8), Geach thinks that the calculus' variables are semantically equivalent to pronouns in natural language. In these respects he holds, as I do not, that the predicate calculus is semantically similar to natural language.

Thirdly, Geach thinks that the use of common nouns – substantival general names – in both the grammatical subject and predicate positions constitutes a systematic ambiguity (ibid., § 88). He suggests (ibid., § 109) analyzing the predicative use 'is an A' of such terms as short for 'is the same A as something', in order to eliminate that predicative use and with it the alleged ambiguity. By contrast, given my minimalist conception of predication – see section 11.5 - I do not think such a double use should be seen as posing any difficulty; consequently, I do not think a reductive analysis of any of these uses is required.

Lastly, unlike Geach, I elaborate my analysis to make it applicable to multiply quantified sentences and sentences containing bound anaphora. And I also develop a deductive system for natural language on the basis of my analyses. Yet despite all these important distinctions, our basic conception of the semantic role of common nouns and quantifiers – which is in substantial agreement with Aristotle's logic – is obviously very close.

Geach concludes a later paper of his (1968) with the following words:

> What we still have not got is a formal theory that recognizes the status of some general terms as names without blurring the distinction between names and predicables. Success in stating such a theory would be Paradise Regained.

Geach sets a surprisingly low standard for Paradise in these lines. Be that as it may, I attempt to state such a theory in Part III below, while maintaining the mentioned distinction. My minimalist conception of predication (see section 11.5), though, and the fact that I allow general terms to function both as predicates and as 'names', might be unacceptable to Geach.

Peter Strawson has also noted the existence of plural referring expressions in natural language, and he has also briefly discussed their relation to quantification. I shall therefore discuss here what he says on the subject. His most elaborate discussion of plural referring expressions is in his book *Subject and Predicate in Logic and Grammar*, in the chapter called 'Substantiation and its Modes'. He maintains there that common nouns are basically predicates, used to specify general characters. They have, however, a secondary, derived use as referring expressions, a function he calls 'substantiation'. (Strawson attributes this function only to countable common nouns, which he calls 'sortals'; but this distinction is irrelevant to my purpose here.) Adjectives and verbs can be used in this way as well, but common nouns are more adequate to that use, he maintains, since they 'signify kinds or sorts of individual substances', and not qualities or types of changes (p. 103).

By contrast, I maintain that the referential use of common nouns is not derived from or secondary to their predicative use. In Chapter 3 I have discussed in some detail the various semantic distinctions between a referential use of concepts and a predicative one. What Strawson sees as common nouns' basic role – i.e., predication – I consider in some cases to be quite distinct from predication by means of adjectives, and I termed it 'classification' in order to emphasize the differences. A sentence such as, say, 'This animal is an elephant', does not attribute any property or general character to the animal referred to, but says what kind of animal it is. By contrast, the sentence 'This animal is dangerous' does attribute a property to the mentioned animal. (Of course, as I have noted in section 3.3, some common nouns, e.g. 'square', can be seen as attributing a property in their predicative use). Thus, what Strawson takes to be a form of substantiation, a secondary function, I take to be a basic function of these expressions, while I draw an important distinction between kinds of predication which he considers as basically of a kind.

Strawson mentions quantification too, and shows what form it would take if plural referring expressions were admitted (1974, pp. 112-3). However, what he says there is very brief and it is left undetermined whether he thinks that in, say, 'Some horses are brown' we refer to some horses and say that they are brown, or to horses in general and say that some of them are brown. In an earlier work (1952, chapter 6, section 8) he seems to maintain the first option, but his treatment of the subject there is also very brief. While the first option, as Geach has shown, is unacceptable, the second one is the one maintained in this work.

6.4   Binary and Restricted Quantification, and Comparative Quantifiers

The standard version of the predicate calculus uses only two quantifiers, 'all' and 'some'. Natural language, on the other hand, makes use of many more quantifiers: 'most', 'many', 'few', 'seven', 'more than seven', 'an even number of', 'infinitely many', etc. All these function in natural language as indefinite determiners; they seem to answer the same semantic principles, and have the same or similar syntax. If the predicate calculus is to be used to analyze the semantics of natural language, it should be able to translate sentences that make use of such quantifiers. Otherwise, it would be committed to the implausible claim that the mentioned similarity is misleading, deep semantic differences being hidden behind insignificant syntactic resemblance. Can these quantifiers be introduced into the predicate calculus?

Some obviously can. Consider, for instance, the numerical quantifier 'two'. We can define this numerical quantifier in the calculus thus: '(Two x)Fx' is true if and only if two particulars in the domain are F. 'Two men came to work' can then be translated as '(Two x)(Man x & Came-to-work x)'.[1] The translation into the predicate calculus of what is said in natural language by means of numerical quantifiers ('two', 'more than two', 'at least two', etc.) does not create any new difficulty.

But this is not the case with all the quantifiers mentioned above. Let us consider the quantifier 'most'. One might have thought that we could introduce a new quantifier into the predicate calculus, 'most', defined as follows·

> For any sentential function 'F', '(most x)(Fx)' is true if and
> only if most particulars in the domain are F.

This would perhaps allow us to translate, for instance, the sentence 'Most things are perishable', but we would still be unable to translate into the calculus the sentence 'Most men are immortal'. If one attempted to translate this sentence as

> (Most x)(Man $x \to$ Immortal x),

---

[1] This quantifier is also definable by means of 'all' and 'some', '(Two x)Fx' being synonymous with '(There is an x)(there is a y)(Fx & Fy & $x{\neq}y$ & (all z)($Fz \to (z{=}x$ or $z{=}y$)))'. But the question, whether a concept can be defined by means of other concepts or be reduced to them, is distinct from the question that interests us here, namely, whether it can be introduced into a language with a given syntax. The quantifier 'infinitely many', as in the sentence 'Infinitely many numbers are prime', cannot be defined by means of 'all' and 'some', but can still be introduced into first-order predicate calculus.

then since most things are not men, the antecedent would be false for most values of x, and therefore the implication would be true for most values of x. In consequence, the supposedly translating sentence is true, although the translated sentence is false. This translation is, therefore, inadequate.

In fact, Rescher (1962) has shown that it is impossible to express in the standard first order predicate calculus what we express by means of 'most' in natural language. The case with some other quantifiers – e.g., 'many' and 'few' – is the same.[2]

On the other hand, taking quantifiers to be modifiers of referring nouns explains why 'most' and 'many' are quantifiers of the same semantic family as 'some' and 'all'. A sentence of the form 'Most A's are B's' is true if and only if most A's are B's, just as 'All A's are B's' is true if and only if all A's are B's. We use 'A' to denote all relevant A's, and 'most' determines to how many of them the predication should apply, i.e., how many are B's. In general, a sentence of the form 'q A's are B's', where 'q' is a quantifier, is true if and only q of the A's are B's. The fact that my analysis of the functioning of common nouns and quantifiers can explain why 'most' and 'some' behave similarly in natural language, while the standard version of the predicate calculus cannot capture this similarity, supports the claim that my analysis of their functioning is more correct than that of this version of the calculus.

The calculus departs, in its treatment of quantifiers, from the way they function in natural language. This departure is partly a result of the fact that the referring expressions admitted by the calculus differ from those admitted by natural language. Because of the latter difference, Frege could not make quantification in the calculus function in the way it does in natural language, and then the option of treating quantifiers as second order concepts suggested itself to him. Now quantifiers like 'most' and 'many' were ignored by Aristotelian logic (although syllogisms involving 'most' were already mentioned by De Morgan (1847, p. 163)), and are inessential or even useless in mathematics. This is probably why the inability of his calculus to treat such quantifiers did not disturb, and perhaps even escaped, Frege.

From the mid-seventies on, following the work of Richard Montague (1973), linguists tried to analyze the semantics of natural language by means of the predicate calculus. At the same time, due mainly to Donald Davidson's influence, philosophers' interest in such an analysis increased as well. The inability of the standard version of the calculus to translate sentences that make use of 'most', 'many' and other quantifiers was therefore acknowledged as a problem. Consequently, modifications of the syntax of quantified sentences of the predicate calculus were suggested, with the intention of extending the calculus' quantificational power. As a result, the quantifiers of natural language are usually construed in contemporary linguistics and logic as either binary or restricted quantifiers. (Binary and restricted quantifiers are logically equivalent,

---

[2] See also Kolaitis andVäänänen, 1995 (Reference according to Westerståhl, 2001).

although some additional constraints – to be discussed below – are usually imposed on restricted quantifiers.[3,4])

Binary quantifiers are quantifiers that operate on an ordered pair of sentential functions – instead of operating on a single sentential function, as is the case in the standard version of the calculus. Their syntax is *(q x)(Fx, Gx)*, where '*q*' is any quantifier. In this system the sentence

All men are mortal

is translated as

(all *x*)(Man *x*, Mortal *x*).

This sentence is true if and only if all the particulars which fall under the first concept – i.e., all men – fall under the second concept as well – i.e., are mortal. A quantifier 'most' can be introduced into this system as well, '(most *x*)(Fx, Gx)' being true if and only if most of the particulars which are *F* are also *G*.

Restricted quantification differs from binary quantification only in its syntax. The basic quantified sentence is of the form:

[*q x: Fx*]*Gx*.

For instance, the sentence 'All men are mortal' is translated as

[all *x*: Man *x*](Mortal *x*).

As with binary quantification, this sentence is true if and only if all the particulars that fall under the restricting sentential function, 'Man *x*', fall under the second sentential function as well. In this modified calculus quantifiers like 'most' and 'many' can also be introduced.

---

[3] Andrzej Mostowski generalized in 1957 Frege's conception of quantifiers, and his work was subsequently developed by other mathematicians. His generalized quantifiers, of which binary and restricted quantifiers are a special kind, established themselves in linguistics mainly following the publication in 1981 of the paper 'Generalized Quantifiers and Natural Language', by Jon Barwise and Robin Copper. Barwise and Copper used restricted quantifiers.

David Wiggins, relying on a short discussion by Frege (1884, § 47), developed during the seventies binary quantification in response to the problem created by 'most'. Although he published his work only in 1981, some philosophers were influenced by it already in the late seventies (Platts, 1979, 100-106; Peacocke, 1979). Similar ideas were simultaneously developed by Gareth Evans (1977b), perhaps independently of Wiggins.

Wiggins was not directly influenced by the developments originating with Mostowski (see Wiggins, 1981, note 24). He was, however, familiar with Evans's work, who refers to one of the first works which applied Mostowski's general quantifiers in linguistics (Altham and Tennant, 1975). Barwise and Cooper, in their turn, were familiar with Peacocke's 1979 paper. Thus the two approaches may not have developed entirely independently of each other.

[4] In my exposition of contemporary theory in linguistics I rely mainly on Keenan and Westerståhl (1997) and on Westerståhl (2001).

Although the problem of translating sentences that make use of 'most', 'many' and some other quantifiers is solved by these approaches, other difficulties confront them, difficulties of which the interpretation of quantifiers as modifiers of referring nouns is free.

Both the binary approach and the restricted-quantification approach can accommodate the quantifiers 'more', 'less', 'the same number of', 'twice as many', 'a larger fraction of', etc. – quantifiers which I shall call *comparative*. For instance, in the binary system, we can define a quantifier 'more' thus:

'(more x)(Fx, Gx)' is true if and only if there are more F's than G's.

In restricted quantification, the sentence

[more x: Fx]Gx

will have the same truth-conditions as the above. These sentences thus translate the natural language sentence

There are more F's than G's.

Similarly, the sentence

More boys than girls smoke

will be translated by binary quantifiers as

(More x)(Boy x & Smokes x, Girl x & Smokes x).

One can thus define binary and restricted quantifiers synonymous with 'more' and other comparative quantifiers of natural language.

Semanticists claim that these modified calculi reveal the logical structure of natural language. The quantifiers of natural language, they claim, are, say, binary quantifiers. In addition, we saw that binary quantifiers synonymous with the comparative quantifiers of natural language are possible. It would thus be natural to expect comparative quantifiers in natural language to answer syntactic principles similar to those answered by quantifiers of the 'all' family. However, the syntax in natural language of the first group of quantifiers is markedly different from that of the second. 'All'-quantifiers complete a *single* noun into a noun phrase, while comparative quantifiers complete *two* nouns into a noun phrase. The former are unary determiners, while the latter are binary determiners. Compare the noun phrases 'all men', 'many men' and 'seven men' with the noun phrases 'more men than women' and 'twice as many men than women'. Typical sentences employing the first quantifiers are

There were seven men at the party

and

Twelve boys came to the theater.

While sentences employing the second are

There were more women than men at the party,

and

More boys than girls came to the theater.

Comparative quantifiers form – not only in English – a syntactically distinct family. If we do not want to commit ourselves to the implausible claim that there is no significant semantic distinction behind these distinct structures, then semanticists should explain why, despite the fact that comparative quantifiers could have been incorporated into the existing binary quantification structure of 'all'-quantifiers, they are not.

The way in which the difference between 'all'-quantifiers and comparative quantifiers in natural language is usually explained in modern semantic theory is by saying that 'all'-quantifiers should be *conservative*. A binary quantifier is conservative in case the quantity of particulars which are B but not A does not matter to the truth-value of the sentence '$(q\ x)(Ax,\ Bx)$'. 'More' is not conservative, since the number of particulars which are B but not A is relevant for determining whether there are more A's than B's; and that applies to all other comparative quantifiers as well. Another constraint on 'all'-quantifiers, needed to avoid some other inappropriate quantifiers, is that they should observe a condition called *extension*: the quantity of particulars in the domain which are neither A nor B should not affect the truth-value of '$(q\ x)(Ax,\ Bx)$'. We can count as *restricted* quantifiers only binary quantifiers which obey these two constraints, and in this way justify the syntactic asymmetry between the two predicates in the symbolism of restricted quantifiers.

But this approach still leaves unanswered the question we started with, namely, why does the syntax of comparative quantifiers differ from that of 'all'-quantifiers in natural language? Natural language does have, according to this approach, binary quantifiers which are not conservative – i.e., 'more' and other comparative quantifiers. Thus, according to this approach, natural language makes a logically unnecessary syntactic distinction between conservative and non-conservative binary quantifiers. The fact that some binary quantifiers are conservative while some are not is an insufficient reason for distinguishing between them syntactically, as the symbolism of binary quantifiers demonstrates. This approach identifies the alleged criteria according to which the syntactic distinction is made, but it does not explain why the distinction should be made.

This discrepancy between the classification of the two quantifier families as forming two separate groups in natural language, and as forming a single family in the modified calculi, indicates that the latter also deviate in their treatment of quantification from the way it functions in natural language.

By contrast, the difference between the two quantifier families is easily explained on the principles argued for in this work. 'All'-quantifiers operate on a single referring expression, determining to how many of the particulars it designates a predicate should apply. Comparative quantifiers operate on *two* referring expressions, determining the relation between the numbers of particulars referred to by the two expressions to which a predicate applies. For instance, 'all', in 'All men are mortal', is used to specify to how many of the particulars referred to by the noun – i.e., to how many men – the predicate 'mortal' should apply; while 'more', in 'More boys than girls smoke', compares the numbers of the particulars referred to by two different nouns to which the predicate applies.

The predicate calculus replaces referring concepts by predicates. As a result, when quantifiers are modified so that they operate on two predicates, the modified calculus cannot distinguish between (i) quantifiers of natural language that operate on a single referring expression and determine to how many of the particulars to which it refers a predicate applies; and between (ii) quantifiers that compare the number of particulars referred to by two expressions. This is why the modified versions of the predicate calculus cannot distinguish between quantifiers of these two families.

On the other hand, the greater success of these modified versions, compared with that of the standard version of the calculus, in translating sentences of natural language, is also explained by the principles presented in this work. The fact that the quantifier, in the binary and restricted versions, operates on two predicates, makes it possible to use the first predicate actually as a distorted referring expression and only the second as really a predicate. In this way 'all'-quantifiers can be translated into the modified calculi.

The analysis of quantification in natural language by means of restricted quantifiers seemed plausible when it was introduced into linguistics only because logicians and linguists were in the grip of the predicate calculus's conception of quantification. When Barwise and Cooper introduced restricted quantification (1981, §§ 1.1-1.2, pp. 160-1), they claimed to 'abstract out the quantifiers at work' in some sentences. For instance, the abstraction of the form of the sentence

1   Most people voted for Carter,

was claimed to be either

Most x such that $\psi(x)$ satisfy $\varphi(x)$,

or, more symbolically,

(most ψ)$x[\varphi(x)]$.

But these two formulas can be considered abstractions from (1) only if we presuppose that quantification involves variables, and that the noun 'people' in (1) is used as a predicate. These two presuppositions – which are, I claimed, mistaken – would have at least been far from obvious unless one was already in the grip of the predicate calculus. These preconceptions are also attested by Barwise and Copper's assumption that quantification presupposes a domain over which variables range. Moreover, their analysis of (1) brought them to claim that the quantifier in that sentence is 'most people', and not 'most' – although they admitted that they were 'at some pains' not to call 'most' a quantifier (ibid., § 1.3). This unnatural convention, which was presented as a discovery (and which was accepted by linguists), attests again to the grip that the predicate calculus's conception of quantification has on logicians. And by the time the authors claim (e.g., in the title of section 1.6), on the basis of their analysis of quantification, that 'proper names and other noun phrases are natural language quantifiers', 'quantifier' has all but lost its connection with quantities.

My analysis of the functioning of nouns and quantifiers has succeeded in explaining the following: Firstly, why 'most', 'many', and other quantifiers that cannot be incorporated into the standard version of the predicate calculus are apparently of the same semantic family as those quantifiers that can be so translated. Secondly, why 'all'-quantifiers and 'more'-quantifiers form two separate families in natural language. Thirdly, why 'all'-quantifiers can be translated into the binary and restricted modifications of the calculus. Finally, why those modifications do not distinguish between the two mentioned quantifier families. All this obviously supports my analysis.

The predicate calculus with binary or restricted quantifiers has greater success than the calculus's standard version in translating quantified sentences of natural language. I nevertheless chose the latter version of the calculus, the original and most widely used one, as the basic target of my criticisms, because the mistake it embodies, the analysis of referring concepts as predicates, is preserved in the modified versions. They are more complex versions which seem more adequate although preserving the same mistake. Accordingly, if my discussion in the first part of this work succeeded in supporting the view that some concepts are sometimes used not as predicates but as referring expressions, these latter versions should be rejected as well if they are claimed to capture the semantics of natural language.

Is 'more' a quantifier at all? This question is misleading. It makes us think that quantifiers have some nature or essence that is independent of our conventions, and that we have to reveal that essence in order to *discover* whether any given word is a quantifier. But this is not the nature of the case confronting us. We called two words, 'all' and 'some', quantifiers, and now we have to *decide* how to expand the concept to new cases, what we would like to include in the concept's extension and what not. This conventional aspect of our procedure doesn't make it arbitrary; conventions are not, as a rule, arbitrary. But the nature of the reasons one should expect is different, and controversies, if any, would be over what is useful, not over what is correct.

Since 'many', 'seven', etc. have many syntactic and semantic characteristics in common with 'all' and 'some', it was most natural to classify them as quantifiers. 'More', 'less', 'twice as many', etc., on the other hand, function quite differently and have a different syntax from all these quantifiers: they are binary, and not unary, determiners. They are also used to construct comparative predicates – e.g., 'more intelligent than' – a use that has no parallel with paradigmatic quantifiers. However, they too are used to determine quantities, and they too are purely formal concepts (their rule of use does not mention any specific property). There are therefore reasons for and against classifying them as quantifiers. I chose to classify them as such, but a different decision would also be acceptable.

6.5   Is 'Existence' a Quantifier?

Aristotelian Logic distinguished between 'some' – the *particular* quantifier – and the concept of existence. For instance, its classification of the valid inferences in which the particular quantifier is involved was not considered as being, among other things, an analysis of the logic of the concept of existence. In fact, an analysis of that concept was never among the canonical parts, or even a central topic of Aristotelian Logic.

By contrast, Frege analyzed particular quantification as a kind of an existential construction. Already in his *Begriffsschrift*, the sentences 'For some x, Fx' and 'There is an x such that Fx' are logically equivalent. The concept of existence was analyzed as a kind of quantifier, the *existential* quantifier, and the particular and existential quantifiers were merged in his calculus. This is of course legitimate, if we regard the calculus as a language with a semantics that does not necessarily reflect that of natural language. This attitude, however, is not the prevalent one. The concept of existence is frequently taken to be such a quantifier in natural language as well, as is attested by the belief that translation into the predicate calculus reveals the logical form of the translated sentences.

I think this is, on the whole, a mistake. In this section I shall try to show that the particular quantifier and the concept of existence should be given different logical analysis, and that we do not even have a sufficient reason to classify the concept of existence as a quantifier. Accordingly, the deductive system I develop in Part III is meant to apply only to the particular quantifier.

How should we define a quantifier in natural language? As I claimed in the previous section, this question is misleading. We are not looking for an essence independent of our conventions; rather, relying on some paradigms, we are trying to devise a natural and convenient concept.

Quantifiers, according to their name, should indicate quantities – they answer the question 'How many?'. And two paradigmatic quantifiers are the indefinite determiners 'all' and 'some'. It is thus natural to classify other indefinite determiners – e.g., 'many', 'most', 'seven', 'at least six' – as quantifiers. These concepts obey the same syntactic and semantic principles as 'all' and 'some'.

The concept of existence, on the other hand, differs from these quantifiers in several respects. Firstly, it is not a determiner at all. We say 'There are horses in the stable', or 'There are many horses', and in both cases existence is not asserted by means of any determiner. We can also say 'King Alfred existed, king Arthur didn't', or 'Some of Homer's heroes existed', where the concept of existence functions grammatically and logically as a predicate (see Strawson, 1967). If we substituted acknowledged quantifiers, such as 'some' or 'seven', for 'there are' or 'existed' in these examples, we would get ungrammatical sentences. Moreover, we cannot substitute 'there are' or 'exist', no matter how we conjugate them, for the quantifiers in, say, '*Some* students were late' or 'There are *many* philosophers' and get grammatical or meaningful sentences. Lastly, the concept of existence cannot be seen as answering the question 'How many?'. The way existence functions in language is very different from the way quantifiers function in it.

These considerations might seem as merely devising apparent differences. Is not the concept of existence used to say that at least one thing has a certain property? If so, then despite its syntactic peculiarity, it does function as a quantifier. Moreover, isn't there logical equivalence between constructions involving 'there is' or 'there are' and those involving the particular quantifier? For instance, 'There are brown horses' is equivalent to 'Some horses are brown'. And logical equivalence, in the sense of mutual entailment, means identity of meaning.

I think that on closer examination the claimed equivalence is revealed as merely apparent. Consider, first, the sentence

1   There are horses.

By contrast to 'There are brown horses', which seems to be equivalent to 'Some horses are brown', this sentence does not have a similar natural parallel. A suggestion one comes across,

2   Some things are horses,

is problematic for several reasons. Firstly, it sounds artificial as a sentence of natural language, and we are considering equivalence in natural language. Secondly, the reference to 'things' is strange: which things are meant? Animals? Or all physical objects? Or perhaps even abstract entities, such as ideas and propositions? Reference to things in general has no determinable boundaries, not even vague ones, and is therefore problematic. Lastly, no mention of 'things' is made in (1), and thus its introduction in (2) violates the claimed equivalence.

But it can be shown that even the standard examples do not demonstrate logical equivalence. Let us examine the two sentences,

3   There are brown horses

and

4   Some horses are brown,

in the case in which there are no horses at all. While (3) is plainly false in that case, the ascription of a truth-value to (4) is problematic. (4) presupposes successful reference to horses, and therefore, if one used it to make a statement in the case we are considering, the right reaction would be to say that there are no horses. But (3) has no such presupposition, and it would thus be appropriate to react to its utterance by simply saying that it is false.

That the claimed equivalence does not hold is perhaps even clearer in the following case. According to their predicate calculus analysis, the following sentences are logically equivalent:

5   There are no brown horses
6   All horses are not brown.

Now in case there are no horses, (5) would be true, while it would be problematic to classify (6) as either true or false. The best reaction to an assertion of (6) in such a case would again be to say that there are no horses; in these circumstances, the ascription of a truth-value to (6) is not the right move in the language-game. By contrast, (5) would be a natural answer to the question, 'How many brown horses are there in the farm?', even if there is no horse in the farm. Moreover, one cannot claim, on Griceian grounds, that (6) is true but misleading, since it implicates that there are horses although it does not presuppose it. If that were the case, then the implicature should be cancelable by locutions such as 'All horses are not brown; but there are no horses' or 'All horses are not brown; but I do not mean to imply that there are any horses' (Grice, 1967, p. 44); but these assertions make no sense.

Thus, the claimed logical equivalence does not hold. The concept of existence and particular quantification, conflated by Frege, are different concepts in natural language. I shall also treat them as distinct. The calculus developed in the next part of this book is supposed to capture the logic of the particular quantifier in natural language, while it ignores the logic of the concept of existence. In fact, if we use 'all' and 'some' as our paradigms of quantifiers, then there isn't even a sufficient reason to classify the concept of existence as a quantifier. The predicate calculus departs from natural language in this respect as well.

Another concept that will not be treated in the deductive system developed in Part III of this book, and which is related in its semantics and logic to the construction 'there is/are', is 'have' in some of its uses. Let us see in what way it differs, when quantification is involved, from other, typical transitive verbs, which are used to express relations. Consider first the sentence

7   John saw three women.

Semantically and logically, 'saw' in sentence (7) expresses a binary relation. On the principles discussed above (and see also section 7.2 below), this sentence is true if and only if 'three women' can be substituted by three singular referring expressions referring to different women so that true sentences will result. For instance, if 'John saw Jane', 'John saw Lydia' and 'John saw Mary' are true, so is 'John saw three women'.

But consider now the sentence

8   John has three sisters.

The same principle does not apply to sentence (8). Suppose that sentence (8) is true, and that one of John's sisters is Jane. If we substitute 'Jane' for 'three sisters', we will get the sentence

9   John has Jane.

But the truth-value of sentence (8) does not relate to that of sentence (9) in the way that the truth-value of sentence (7) related to those of *its* instances. Moreover, (9) is ambiguous in a way that (8) is not: out of a context, what 'has' means in (9) is indeterminate, despite the fact that no such indeterminacy is involved in sentence (8); and no such relative ambiguity was involved in the relation between sentence (7) and its instances. Semantically and logically, this use of 'have' is markedly different from that of verbs that express relations.

This use of 'have' should be distinguished from its use as synonymous with 'own'. What was said above about the relation between the truth-value of sentence (7) and those of its substitution instances does apply, perhaps, to the relation between the truth-value of, say, 'John has three bicycles' and those of its substitution instances. If that is the case, then the deductive system developed in Part III of this book should apply to 'have' in its use as synonymous with 'own'.

The peculiarity of 'have' in its use as in sentence (8) is reflected by the way such sentences are translated into the predicate calculus. Consider the two sentences

10   John saw a woman,
11   John has a sister.

They are usually translated into the first order predicate calculus as, respectively:

12   (There is an x)(Woman x & John saw x)
13   (There is an x)(Sister(x, John)).

While the verb 'saw' reoccurs in (10)'s translation as a two-place predicate, the verb 'has' does not have such a parallel in (11)'s translation. And while 'woman' is a one-place predicate in (10)'s translation, 'sister' is a two-place predicate in (11)'s. Of course, the relation between (10) and (12) is the typical one: 'John φ's a ψ' is generally translated as '(There is an x)(ψx & φ(John, x))'. The verb 'have' in this use is logically exceptional.

By contrast to many other transitive verbs, 'have', as used in sentences (8) and (11), expresses no relation. This can perhaps be made clear if we ask, what relation *does* it express? No answer is forthcoming. And of course, one should not answer that it expresses the relation of *having*: to have a sister, an accident, a good reason for staying at home and a headache are utterly different things. Rather, we use 'have' in sentences like (8) and (11) to say that a relation, specified by the concept of the grammatical object ('sister' in our examples), holds between the subject and someone or something. For instance, in uttering (11), 'John has a sister', we say that the relation of *being a sister of* holds between someone and John; i.e., that someone is a sister of John. Similarly with quantifiers: when we say 'John has *three* sisters', we say that the relation of *being a sister of* holds between three people (or, more generally, particulars) and John.

In this respect 'have' resembles 'there is', since the latter is used to say that something, specified by 'α' in 'There is an α', holds of- or is the case with someone or something.

The semantic affinity of 'have' and 'there is' is more apparent when we consider their use with some three-place predicates, instead of two-place ones. For instance, 'a is between b and c' (instead of 'a is a sister of b'). The following two sentences are synonymous:

London and Oxford *have* two cities between them.
There are two cities between London and Oxford.

It is also relevant to note that some languages, e.g. Hebrew, use their translation of 'be' to translate not only English sentences in which 'there is' is used, but also those mentioned above in which 'have' is used.

The logic of 'have' in its use that was considered above clearly has more to it than what was indicated here. But for our discussion it is sufficient if we have shown that 'have' in this use is such a peculiar verb that a logic of relations can consider it an exception. And this will indeed be my approach in Part III below.

Chapter 7

Multiple Quantification

7.1   On Ambiguity and Formalization

The subject of this chapter is multiple quantification. More specifically, we shall investigate the logic and semantics of sentences in which a many-place predicate has two or more quantified noun phrases among its subjects. For instance: 'Every man loves several women' or 'Three girls bought seven pencils each'.

If we intend not only to criticize the adequacy of the predicate calculus for the analysis of natural language, but also to suggest an alternative, then this subject is of special importance for us. It is commonly and rightly claimed that the predicate calculus is a major advance on Aristotelian logic in its treatment of the inference relations between sentences that involve multiple quantification (e.g., Dummett, 1981, p. xxxii). While Aristotelian logic did not and could not handle an inference like

> Some women are loved by every man; Hence, every man
> loves some women,

the validity of its translation into the predicate calculus is easily established. Any alternative logic should have comparable power.

As preliminaries I shall discuss in this section two topics, ambiguity and formalization.

I start with ambiguity. Sometimes, sentences of the predicate calculus that involve multiple quantification are not ambiguous while the sentences of natural language that they translate are. Although the ambiguity of natural language is often exaggerated, some sentences with multiple quantification do admit of several readings. 'A man went into every store', for instance, can mean either that there is a certain man who went into all stores; or that into every store went a man, possibly different men into different stores (Higginbotham's example). Translating this sentence into the predicate calculus forces us to disambiguate it: each meaning is translated by a different sentence.

Of course, we can disambiguate the sentence within natural language as well – this is, in fact, what I have just done when I explained the ambiguity. The context of utterance or the meaning of some of the words in the sentence are usually sufficient to make clear which reading of the ambiguous sentence the speaker has in mind.[1]

---

[1] When I say that the meaning of some of the words in the sentence is sometimes sufficient to disambiguate it, I have in mind sentences like 'An oak grew from every acorn' (Jackendoff's

Speakers therefore often naturally use ambiguous sentences. But natural language is not inescapably ambiguous. We can express ourselves more lengthily and avoid ambiguity, as we do when we think that misunderstanding might or did arise.

Yet our logic should apply not only to these longer and non-ambiguous sentences, but to our common sentences as well, sentences which *are* often ambiguous. But ambiguity does not make a systematic treatment of inferential relations between sentences impossible. It does, however, compel one to specify the intended reading of the sentences to which the system is supposed to apply. In the following sections I shall therefore distinguish between principles that give distinct readings of multiply quantified sentences, and then supply the required specification.

I proceed to discuss formalization. What is often called 'formalization' in logic is actually a *translation* into another language, that of the predicate calculus. If to formalize a sentence should mean (as it often does, and as I shall use it below) to write a formula that (i) abstracts from the content of the words in the sentence and (ii) makes clear their function in the sentence and the relations between them, then the formalization of 'Every whale is a mammal' is 'Every *A* is a *B*'. The sentence '(every *x*)(Whale *x* → Mammal *x*)' is *not* a formalization, in *this* sense, but a *translation* of the original into the predicate calculus. It cannot be a formalization, since it does not abstract from the content of the words in the sentence it translates. The fact that single letters are often used in translating English words into the predicate calculus helps generate the illusion that the translation is a formalization. And to call the sentence-form, generated by substituting predicate-variables for the concepts in that translation, a formalization, is to assume that the concepts in it function in the same way and stand in the same relations as those in the translated sentence. But if what I have argued in this book is correct, this assumption is mistaken.

I have just written that the logical form of 'Every whale is a mammal' is given by the formula 'Every *A* is a *B*'. This is correct, but there is more to formalization in logic than that. Consider the sentences

> I am tall,
> Paul is tall.

Suppose we formalize these sentences by merely substituting variables for concepts and for referring expressions. We can substitute variables of one kind – say *a*, *b*, etc. – for singular referring expressions, and of another kind – say *A*, *B*, etc. – for adjectives. We then get the two formulas:

---

example), which we read only as meaning that a different oak grew from each acorn. We ignore the other reading, i.e., that one and the same oak grew from all acorns, because we know it describes an impossibility. However, if the alternative reading had described something possible and the speaker had wanted to avoid misunderstanding, the speaker could have said 'From every acorn grew an oak', which does not allow of the alternative reading.

> *a* am *A*,
> *a* is *A*.

These formulas are different, although the two formalized sentences have the same logical form, at least on some level of abstraction. Our formulas preserve merely grammatical distinctions (which add constraints on possible substitutions – 'I' can be substituted for '*a*' in the first formula, but not in the second one). If we want to be capable of expressing their identical logical form in a formula, we need to substitute a variable for the copulas in the two sentences. For instance, we can use 'is' as a *variable*, to be substituted in accordance with the noun which is substituted for *a*. The form of the two sentences is then given by '*a* is *A*'.

Complications of this sort often make it convenient to devise an artificial language. Every language – artificial as well as natural – has its own grammatical peculiarities, but perhaps we can devise an artificial language in which sentences which have the same logical form will not differ grammatically. A language which has the same copula for first, second and third person, singular or plural, will overcome the complication we have just met while formalizing English sentences. Such a language may be found more convenient than English for formal proofs.

But devising an artificial language is a delicate matter. If an artificial language is intended as a tool for investigating the logical properties of a natural language, it should be semantically isomorphic (see p. 17) to the latter. Yet a mistaken semantic analysis might cause one wrongly to assume that the desired isomorphism has been achieved. That has been the case, as I have argued, with the predicate calculus.

Moreover, an artificial language may be semantically isomorphic to natural language in some respects only. It may be semantically isomorphic to natural language in the relation between subject and predicate in non-modal sentences, but incapable of such an isomorphism in modal ones, say. The success of the artificial language in the former respect might then mislead us into thinking that its way of incorporating modality also parallels that of natural language.

It is therefore prudent to use artificial languages sparingly. Accordingly, my formalization below is mainly within English. All the same, I found it convenient to devise an artificial language in order to formulate some semantic and logical rules. That language will be gradually introduced below. It is intended only for investigating the semantic and logical properties discussed in this work. It might very well be incapable of capturing other semantic and logical properties of natural language.

7.2   Iterative Reading of Multiply Quantified Sentences

The question I shall discuss in this and the following section is: how does the meaning of a multiply quantified sentence relate to that of a sentence with a single quantified noun phrase? I said above that a sentence of the form '*q A*'s are *B*'s', where '*q*' is a quantifier and '*q A*'s' a noun phrase, is true if and only *q* of the particulars designated by '*A*' are *B*'s. More generally, suppose a sentence contains an *n*-place predicate with *n* noun phrases as subjects (and I shall discuss only such sentences in this chapter), and that only one of these noun phrases is quantified – i.e., is of the form '*q A*'. (For instance, 'John loves several women' or 'Mark gave Jane two presents'.) Then that sentence is true if and only if *q* definite singular noun phrases, each referring to a different *A*, will generate a true sentence if substituted for '*q A*'. The question I shall now address is, how does this rule generalize to sentences with several quantified noun phrases?

As we shall see, there isn't a single way in which this rule is generalized to multiply quantified sentences. I shall discuss one way in which it is generalized in this section, and some additional ways in the following one.

Let us start by examining an example. Consider the sentence:

1   Most women are loved by some men.

I think the most natural way, perhaps the only way of understanding this sentence, is that it is true if, e.g., Jane is loved by some men, Mary is loved by some men, and so on for most women – where different men may love different women. As can be seen, in explaining the meaning of sentence (1), I applied the above substitution rule to the first noun phrase appearing in it: sentence (1) is true if and only if 'most women' can be substituted by names of most women, each substitution generating a true sentence. In this way many sentences are generated – for instance, 'Jane is loved by some men' – each of which is true if and only if 'some men' can be substituted by names of some men, each substitution generating a true sentence. The existence of a second quantified noun phrase in sentence (1) was irrelevant when the substitution of names of women for 'most women' was concerned; similarly, when we discussed substitutions of names of some men for 'some men' in 'Jane is loved by some men', the fact that 'Jane' was substituted for 'most women' was again irrelevant.

We thus see that one way in which the substitution rule applicable to sentences with a single quantified noun phrase is generalized to sentences with several quantified noun phrases is the *iterative* one: we apply the same rule again and again, according to the order in which the quantified noun phrases appear in the sentence. This iterative application parallels the way in which the meaning of multiply quantified sentences in the predicate calculus is defined.

Although my examples above, and also those below, are in English, I have checked the validity of the iterative substitution rule in a wide variety of languages, including a variety of European languages, Chinese, Hebrew, Korean and others. I have always consulted native speakers. In all those very different languages the iterative rule seems to apply. Perhaps this rule is a non-trivial language universal.

The order of the iterative application of this rule can be altered by considerations regarding the plausibility of what is said. Consider, for instance, the sentence 'An oak grew from every acorn'. If we applied our substitution rule according to the order in which the noun phrases appear in this sentence, the meaning that would follow would be absurd: one and the same oak cannot grow from all acorns. We therefore understand this sentence according to a different order of application: an oak grew from this acorn, an oak grew from that acorn, an oak grew from the acorn over there, and so on.

I think it is felt that the understanding of 'An oak grew from every acorn' is influenced by considerations on the plausibility of what is said. If someone were interrupted after having uttered only the first four words of the sentence, 'An oak grew from', we would think he intended to say something about a *single* oak; only if we hear the last part of the sentence as well, 'every acorn', is this understanding ruled out because of its implausibility, and instead we understand the sentence in accord with the rule that substitutes the noun phrases in the inverse order. In addition, the more natural way of saying what that sentence says would be by means of the sentence 'From every acorn grew an oak', whose meaning *is* given by substitutions according to the order in which the noun phrases appear in the sentence.

Accordingly, although some exceptions do occur in language, the iterative application of the substitution rule according to the order of appearance of noun phrases in the sentence is the basic case. The exceptions are not due to the syntax of the sentence concerned, but to considerations of plausibility etc. There is something *ad hoc* in such interpretations, characteristic of the flexibility of natural language. In my discussion below I shall therefore consider only iterative applications of the substitution law according to the order of appearance of noun phrases in a sentence.

I shall now formulate the rule for iterative substitutions more accurately. We shall be concerned with sentences of the form '$(np_1, \ldots np_n)$ *is P*'; that is, syntactically, sentences in which an n-place predicate is predicated of n noun phrases, $n \geq 1$. For instance, 'John loves Mary' ($n=2$) or 'John gave Mary a present' ($n=3$). We shall be interested in cases in which some of the noun phrases 'np_1', $\ldots$ 'np_n' are quantified noun phrases, of the form '[quantifier][general noun]'. Suppose the first quantified noun phrase in '$(np_1, \ldots np_n)$ *is P*' is 'np_i', which is of the form 'qA', where 'q' is a quantifier and 'A' a plural referring expression. Then '$(np_1, \ldots np_n)$ *is P*' is true if and only if '$(np_1, \ldots a, \ldots np_n)$ *is P*' is true for q 'a's, where 'a' is any singular referring noun phrase which has been substituted for 'np_i', 'a' refers to a particular to which 'A' refers, no two 'a's refer to the same particular, and some 'a' refers to any particular to which 'A' refers.

The above rule is insufficient. It should be generalized to cases involving propositional combination of sentences of the above form, i.e., to sentences compounded by means of truth-functional sentential connectives from sentences of that form. It should also be generalized to cases involving definite noun phrases anaphoric on quantified noun phrases. These generalizations will be the subject of a later section (8.6). I gave this rule here only to demonstrate the kind of substitutional considerations that can be used to explain the relation between the truth-values of sentences with quantified noun phrases and sentences without ones.

As will have been noticed, I have just started developing an artificial language, which I shall use below. That language has quantifier-variables – 'q', 'q_1', 'q_2', etc.; predicate-variables, standing for one- or many-place predicates – 'A', 'B', $\ldots$ 'P', etc.; variables for singular referring expressions – 'a', 'b', $\ldots$ 'a_1', 'a_2', etc.; and noun phrase-variables – 'np_1', 'np_2', etc. A noun phrase is either a singular referring expression or an expression of the form '$q\,A$', where 'A' is a one-place predicate-variable. An atomic sentence of that language is of the form '$(np_1, ... np_n)$ *is P*', where '*P*' is an *n*-place predicate-variable.

I shall use this artificial language only in order to make some general statements about sentences of natural language. I shall therefore need only variables, and not constants, for my artificial language. The only exception will be in my use of some specific quantifiers. It will also be noticed that one-place predicate-variables are used both in the predicate position – i.e., after the copula – and as parts of noun phrases. This is supposed to reflect the use of some concepts both as plural referring expressions and as predicates; e.g., 'philosophers' in 'All philosophers are mortal' and 'Some Athenians are philosophers'.

I shall now give a few examples of the application of the iterative substitution rule, in order to examine its correctness. I have already examined its application to a sentence with two quantified noun phrases, 'Most women are loved by some men'. It is easy to see that it also applies to the sentence 'Some men love most women'. Let us next examine two examples of its application to sentences with three quantified noun phrases and a three-place predicate. The predicate I shall use in my examples is '*a* sent *b* to *c*'.

Consider first the sentence 'Two ministers sent five delegates to several countries'. According to our iterative substitution rule, we first find two singular expressions referring to different ministers that we can substitute for 'two ministers' and have a true sentence, say 'the Minister of Education' and 'the Foreign Secretary'. We thus have the two sentences, 'The Minister of Education sent five delegates to several countries' and 'The Foreign Secretary sent five delegates to several countries'. We then find for each of these two sentences five singular expressions referring to different delegates we can substitute for the second noun phrase, 'five delegates', and get true sentences. For each of the ten sentences we now have we should then find several names of different countries we can substitute for 'several countries' and have true sentences.

I believe this procedure gives *a possible* reading of the sentence we started with, 'Two ministers sent five delegates to several countries'. Consider a situation in which to send a delegate to several countries is considered inappropriate. The government's records are examined, and it is discovered that norms have really deteriorated: two ministers sent *five* delegates to several countries.

Let us next examine the sentence 'Five delegates were sent by two ministers to several countries'. According to our iterative substitution rule, this sentence is true if, for instance, 'John was sent by two ministers to several countries' is true, and similarly for four other names of different delegates. Now 'John was sent by two ministers to several countries' is true according to our rule if we can substitute names of two different ministers for 'two ministers' and get a true sentence; for instance, 'John was sent by the Minister of Education to several countries' and 'John was sent by the Foreign Secretary to several countries'. We can now substitute names of several countries for 'several countries' in each of the ten sentences generated in this way, not necessarily the same names in each sentence.

I believe this procedure again gives a possible reading of the sentence we started with this time, namely 'Five delegates were sent by two ministers to several countries'. Consider now a situation in which a delegate who was sent to more than one country is a candidate for promotion, especially one sent by more than a single minister to several countries. In that situation a statement made by the above sentence would naturally be taken to mean what it should mean according to our iterative substitution rule.

7.3   Additional Readings of Quantified Sentences

I believe my two last examples made it clear that although our iterative substitution rule gives possible readings of sentences, the readings it gives are not always the only possible ones, nor even always the most natural ones. In this section I shall consider some readings which are not according to it. This subject is of special significance for my work, since my systematization of the logic of quantified sentences in Part III of this book is intended to apply only to some of these possible readings.

Let us begin with a reading of quantified sentences that is not according to our substitution rule for a sentence with a *single* quantified noun phrase, the rule formulated at the beginning of the previous section. What I have in mind is the distinction between distributive and collective predication, in its application to quantified sentences. Consider the sentence

Three men lifted the table.

This sentence can mean either that each man lifted the table by himself (distributive predication), or that the three men lifted the table together (collective predication). On the collective reading of this sentence, what is true of the plurality considered together need not be true of each of its constituents. It is at least misleading, and perhaps even wrong, to infer from the fact that John, Bill and Peter lifted the table together, that John lifted the table. The collective meaning can be unambiguously conveyed by adding 'a group', 'a set', 'together' and similar locutions to the sentence. But even without such additions, collective readings are very often most natural.

Predication in all the examples of the previous section was taken to be distributive. If it is allowed to be collective as well, many more readings are possible. Take, for instance, the sentence 'Two ministers sent five delegates to several countries'. It can be read as synonymous with 'A group of two ministers sent five delegates to several countries', where the two ministers acted together, and only five delegates were sent – possibly each delegate to different countries. It can also be read as synonymous with 'A group of two ministers sent a group of five delegates to several countries', where the ministers acted together and sent the five delegates as a group to each of the several countries. When we allow for either distributive or collective predication, some sentences will have several possible readings.

Although our substitution rule, as it stands, does not apply to collective predication, it can be straightforwardly modified for these cases. We considered above a sentence that consists of an n-place predicate applied to n noun phrases as subjects, one of which is of the form '$q\,A$'. For *distributive* predication, this sentence is true if and only if q definite singular noun phrases, each referring to a different A, will generate a true sentence if substituted for '$q\,A$'. For *collective* predication, this sentence is true if and only if a definite noun phrase referring to q different A's will generate a true sentence if substituted for '$q\,A$'. For instance, 'Three men lifted the table' is true because 'John, Bill and Peter lifted the table' is true, and 'John', 'Bill' and 'Peter' designate three different men.

The substitution rule for collective predication gives the correct results for distributive predication as well. (Although there are slight differences between the applications of this rule to different quantifiers; e.g., 'exactly three' versus 'at least three' versus 'at most three'.) 'Three men went to sleep' is true if so is, say, 'John, Bill and Peter went to sleep'. It can therefore be considered as the substitution rule for *any* form of predication, whether collective or distributive; the specific substitution rule for distributive predication is then *derived* from the general substitution rule.

This analysis of the relations between the truth-values of a quantified sentence and those of its instances has the significant advantage of not having to consider as ambiguous the form of quantification involved in quantified sentences that contain both collective and distributive predication. For instance,

Three men lifted the table and then went to sleep

is true since, say, the following is true:

John, Bill and Peter lifted the table and then went to sleep.

I have emphasized above (§ 2.3) how implausible it is to maintain that such sentences involve a semantic ambiguity of the noun phrase. The approach developed here avoids this implausibility.

When we come to the systematization of inferential relations between sentences with quantified noun phrases, the rules of inference we shall formulate will not be intended to apply to collective readings of these sentences. For that reason, despite the more fundamental semantic status of the general substitution rule, I shall continue to use substitution rules that are specific to distributive predication. Distributive readings can be disambiguated from collective ones by inserting 'each' in appropriate places in the sentence. 'Two ministers sent each five delegates, each to several countries' allows only of a distributive reading.

The uniformity of analysis of quantification with collective and distributive predication is among the advantages of the approach developed here over an analysis of some sentences that has been developed by George Boolos (1984). Boolos noted (pp. 57-8) that while the sentence:

1   There is a horse that is faster than Zev and also faster than the sire of any horse that is slower than it

can be translated into first-order logic as follows:

($\exists x$)(x is faster than Zev & $\forall y$[x is faster than y → x is faster than y's sire]);

the following sentence cannot be translated into first-order logic:

2   There are some horses that are faster than Zev and also faster than the sire of any horse that is slower than them.

Taking the locution 'there are some horses' as quantifying over a collection, class or totality of horses, Boolos translated sentence (2) into second-order logic as follows (where 'X' ranges over such collections):

$\exists X$($\exists x \, Xx$ & $\forall x$(Xx → x is faster than Zev) & $\forall y$[$\forall x$(Xx → x is faster than y) → $\forall x$(Xx → x is faster than y's sire)]).

Boolos showed that this translation is not equivalent to any first-order sentence.

I think it is implausible that both of Boolos's translations capture the semantic structure of these sentences. The structure of (1) is so close to that of (2), that a very good reason is required for claiming that they involve very different kinds of quantification: quantification into argument position versus quantification into both argument and predicate position. Boolos found these very different translations acceptable only because he presupposed that the semantics of both sentences should be captured by some version of the predicate calculus, either first-order, second-order, or perhaps some other variation on it.

By contrast, on my approach we can analyze sentences resembling (1) and (2) in a uniform manner. (I say sentences *resembling* (1) and (2), because I do not, here or elsewhere in this book, analyze existential constructions such as 'there is' or 'there are' (see section 6.5).) The following two sentences are very close paraphrases of (1) and (2):

3   Some horse is faster than Zev and also faster than the sire of any horse that is slower than it

4   Some horses are faster than Zev and also faster than the sire of any horse that is slower than them.

Sentence (3) is true if we can substitute the name of *one* horse for 'horse' and get a true sentence. For instance:

Bucephalos is faster than Zev and also faster than the sire of any horse that is slower than it.

While sentence (4) is true – taking 'some horses' to mean 'at least two horses', as Boolos does with 'there are some horses' – if a *conjunction* of names of *at least two* horses can be substituted for 'some horses' and yield a true sentence. For instance:

Bucephalus and Pegasus are faster than Zev and also faster than the sire of any horse that is slower than them.

(The appearance of 'it' and 'them' in sentences (3) and (4) requires an analysis of bound anaphora, which is supplied in section 8.2; but I believe my claims can be appreciated independently of this analysis.)

I shall make here a small digression in order to mention Boolos's discussion of the semantics of second-order logic (ibid., pp. 64-72, and further developed within the framework of a Tarskian truth-theory in (Boolos, 1985)), in the process of which he develops a conception of plural reference similar to mine. In his interpretation of second-order logic, Boolos tries to avoid Russell's paradox involving the set whose members are all and only those sets which are not members of themselves. Wishing to make assertions like '$\exists X \forall x [Xx \leftrightarrow \neg x \in x]$', where '$x$' ranges over all sets, Boolos cannot allow 'X' to range over sets. He therefore develops a semantics similar to Russell's early conception of classes as many, which was developed mainly for the same purpose.[2] He urges us (1984, p. 66) to 'abandon, if one ever had it, the idea that use of plural forms must always be understood to commit one to the existence of sets (or "classes," "collections," or "totalities") of those things to which the corresponding singular forms apply.' And further on he writes that (p. 72)

> a second-order quantifier needn't be taken to be a kind of first-order quantifier in disguise, having items of a special kind, collections, in its range. It is not as though there were two sorts of things in the world, individuals, and collections of them, which our first- and second-order variables, respectively, range over and which our singular and plural forms, respectively, denote. There are, rather, two (at least) different ways of referring to the same things...

Boolos conception of plural reference seems similar to mine. But the differences between us are significant. Firstly, Boolos mentions only plural existential quantification, 'there are', as an example of reference to a plurality; he does not mention plural definite noun phrases, such as plural pronouns, plural demonstratives, or conjunctions of proper names, in any of his examples. Secondly, and most significantly, he analyzes common nouns as predicates, and never as plural referring expressions. Thirdly, and partly as a consequence, his conception of quantification is Frege's, while I argue that Frege's conception is inapplicable to natural language. While Boolos's understanding of plural reference is similar to mine, we incorporate it in semantics and logic in very different ways.

Let us proceed to examine structural ambiguities that involve *multiple* quantification. The sentence

5   Three men loved seven women

---

[2] See Russell, 1903, section 70 and chapter X (especially sections 104-6). Boolos concludes his 1984 paper with an (inaccurate) quotation from section 127 of that work, in which the concept of a class as many is again mentioned, distinguished from that of a class as one, and it is claimed that 'assertions can be made about classes as many, but the subject of such assertions is many, not one only as in other assertions.' I take Boolos's quotation and reference to be an implicit acknowledgment of his debt to Russell.

the same seven women. The sentence can be read as roughly synonymous with 'There were three men and seven women, each man loving each woman'. (I say that these sentences are only *roughly* synonymous, since they differ in their presuppositions: the first presupposes, while the second asserts, that there were men and women.) This last reading is distributive, not collective, since it is meant that *each* of the three men loved *each* of the seven women. On this reading the sentence is synonymous with its passive correlate,

Seven women were loved by three men

(where this is also read as meaning *the same* three men). The sentence is true if there are three men and seven women, such that if we substitute 'three men' by any name of one of these men, and 'seven women' by any name of one of these women, we get a true sentence.[3]

As we see, on such a reading the substitution rule for sentences with a single quantified noun phrase is generalized in a non-iterative way. The several quantified noun phrases are substituted *in parallel*, so to say. For each quantified noun phrase of the form '*q A*', *q* substitutions should be made, each with all substitutions of other quantified noun phrases in the sentence. The relation we observed in singly quantified sentences of a quantified noun phrase to its substitution instances is preserved, but these substitutions relate in a non-iterative way to substitution instances of other noun phrase.

We can also consider the meaning of sentence (5) as generated by an *iterative* application of the substitution rule for *collective* predication, which I claimed could be seen as applying, in the case of a single quantified noun phrase, to distributive predication as well. In case more then a single quantified noun phrase is contained in the sentence, this rule can be generalized in two ways. The distribution of the predicate can be performed either after each substitution of a definite noun phrase for a quantified one is made, or after all substitutions have been made. In the first case, the resulting meaning would be identical to the one generated by *iterative* applications of the substitution rule for distributive predication; in the second, the generated meaning would be identical to that generated by a *parallel* application of that rule. Suppose, for instance, that sentence (5) is true because it yields a true sentence when 'three men' is substituted by 'Peter, John and Harry':

Peter, John and Harry loved seven women.

We can still understand this sentence in two ways. Either perform the distribution now, and get the result that Peter loved seven women, and so did John and Harry. This is the meaning identical with the one resulting from the iterative application of the distributive substitution rule. Or first apply the substitution rule again, say substituting 'Paul's seven sisters' for 'seven women', and only then distribute the predicate, getting the result that Peter loved Paul's seven sisters, as did John and

---

[3] The translation of such sentences into the predicate calculus is usually done by introducing *branching* quantifiers. See Barwise, 1979, section 2.

Harry. This meaning is identical with meaning resulting from the *parallel* application of the distributive substitution rule.

The reading of multiply quantified sentences which parallels multiple quantification in first order predicate calculus is the distributive-iterative one. My logical system in the next part of this work, which is meant to parallel that of the predicate calculus, applies only to the distributive-iterative reading of multiply quantified sentences.

Let us examine another possible reading of multiply quantified sentences. Consider again the sentence

6   Three men loved seven women.

As we saw, it allows both of an iterative-substitution reading and of a parallel-substitution one. However, it has another possible reading: perhaps one man loved, say, two women, another man loved four women, and a third man loved a single women, the three men loving *altogether* seven women. That is, the sentence can be used as synonymous with 'Three men loved altogether seven women'.

In this case the substitution rule for singly quantified sentences generalizes in a *cumulative* way to multiply quantified sentences. The number of different substitutions of the second noun phrase, 'seven women', that yield true sentences for each substitution of the first noun phrase, 'three men', should add up to seven. This reading can thus be called *cumulative*.

By contrast to the former cases we considered, it is not equally straightforward to see the cumulative reading as resulting from some sort of application of the collective substitution rule.

We see that multiply quantified sentences allow of various kinds of readings, which would be natural in various contexts and for various concepts. These are the distributive against the collective readings, and the iterative, parallel and cumulative readings. They can be at least partially disambiguated by adding locutions like 'each', 'a group of', 'altogether', etc. A logical analysis should always note which reading is being considered. The logical system I develop in the third part of this work is meant to parallel the first order predicate calculus, and it applies only to the distributive-cum-iterative reading.

The iterative, parallel and cumulative readings were distinguished by their different relations to sentences obtained by substituting singular definite noun phrases for quantified noun phrases. Whenever seven women were mentioned, we had to substitute names of seven women for 'seven women'; but the ways these substitutions were made vis-à-vis other substitutions in the same sentence varied. Accordingly, approaching quantification through its relation to substitution enables us to see how quantified noun phrases make a uniform contribution to meaning, although their contribution can be incorporated in various ways in sentences.[4]

---

[4] My examples of different ways of generalizing the substitution rule for multiply quantified sentences are not meant to be exhaustive. There are multiply quantified sentences that cannot

7.4   On the Passive, Converse Relation-Names, and the Copula

As can be seen from our iterative substitution rule, in order to have the ability to express all possibilities involving quantities of different kinds of things that stand in a given relation, a language should be capable of rearranging in all possible orders the noun phrases contained in a sentence attributing such a relation. Different languages may accomplish this by different means. In English, prepositions and the passive and active voices are the main means for that purpose. To continue the above example, English allows of the following six sentences:

> Two ministers sent five delegates to several countries.
> Two ministers sent, to several countries, five delegates.
> Five delegates were sent by two ministers to several countries.
> Five delegates were sent to several countries by two ministers.
> To several countries five delegates were sent by two ministers.
> To several countries two ministers sent five delegates.

These sentences, when read according to the above iterative substitution rule, describe six different possibilities.

Here lies the real importance of the passive voice. The passive is frequently rather artificial when used without quantified noun phrases. For instance, to say 'Mary is loved by John' instead of 'John loves Mary' may achieve some different emphasis, but the use of the passive instead of the active voice does not involve any change of meaning in this case. Frege, generalizing from such examples, thought that transitions from active to passive and *vice versa* 'can always be made *salva veritate*', and that only 'stylistic and aesthetic reasons' give preference to any form. According to him, 'the distinction between the active and passive voice' is among those made 'from a purely psychological point of view', distinctions which must be rejected in logic (1897, pp. 141-3). Indeed, his predicate calculus contains nothing of the sort.

However, *pace* Frege, the passive is essential for the expressive completeness of English when quantified noun phrases are involved. That is why a passive sentence like, e.g., 'Most women are loved by some men' does not sound artificial. The active form, 'Some men love most women', has a different meaning. The use of the passive voice is essential in this case, if we want to say what the sentence says. (Natural language rarely uses dubious 'Calculish' sentences, hybrids of English and the predicate calculus, like 'For most women it is the case that some men love them'; it prefers to use combinations, by means of sentence-connectives, of subjects-predicate sentences.)

---

be understood according to the generalizations I discussed, e.g., 'Every man loved a different woman'. Some additional examples can be found in Keenan (1992) and Keenan and Westerståhl (1997, § 2.2) (although I think some of their examples are not of additional ways of understanding multiply quantified sentences). At least some of the additional principles according to which these examples are understood can also be seen as additional generalizations of my substitution rule, as I believe is clear of the example I just gave.

Similar considerations clarify the semantic necessity of the existence both of relation-names and their converses in natural language. A relation-name '*R̆*' is the converse relation-name of a relation-name '*R*' if and only if, for any two particulars *a* and *b*, *aRb* if and only if *bR̆a*. For instance, 'teacher of' is the converse relation-name of 'pupil of', since Aristotle is Alexander's teacher if and only if Alexander is Aristotle's pupil. Now it might seem that if a language contains a certain relation-name, then its converse is, in a sense, redundant: it does not add to what can be said by that language. Simply say that *b* is *a*'s pupil instead of saying that *a* is *b*'s teacher, and you can eliminate 'teacher' from your vocabulary. In reality, there is only one relation corresponding both to 'teacher of' and to 'pupil of'; therefore one relation-name should be sufficient.

That this is not so is made clear by considering the following example:

Some lecturers are teachers of every student.

An attempt to say the same thing by means of 'pupil of' instead of 'teacher of' would fail (unless we again countenance Calculish sentences). This failure is explained by our iterative substitution rule. In order to describe every situation in which more than a single particular stand in a given relation, a language has to be able to arrange the expressions designating the related particulars in any order. For that purpose, it has to have two expressions designating the same relation: a relation-name and its converse. Relation-names and their converses are therefore essential for the expressive completeness of language.

The semantic necessity of converse relation-names for natural language demonstrates again the inadequacy of the predicate calculus, even in its versions that use generalized quantifiers, for the analysis of natural language. Suppose we translate '*a* is a pupil of *b*' as 'Pupil(*a*, *b*)'. Then, using restricted quantifiers, 'Some lecturers are teachers of every student' and 'Every Student is a pupil of some lecturers' translate as '[Some *x*: Lecturer *x*][Every *y*: Student *y*]Pupil(*y*, *x*)' and '[Every *y*: Student *y*][Some *x*: Lecturer *x*]Pupil(*y*, *x*)', respectively. There is no semantic need for both a relation-name and its converse for this artificial language. This indicates that this language is semantically different from natural language.

Our iterative substitution rule for natural language also explains the necessity of a copula, or some similar device, for natural language. In the predicate calculus, sentence negation can either precede or follow a quantifier, thus generating two sentences with different meanings. For instance (I use restricted quantification to demonstrate these distinctions),

[Some *x*: Man *x*]Not(Greek *x*)

and

Not[some *x*: Man *x*](Greek *x*)

mean, correspondingly, that some men are not Greek, and that it is not the case that some men are Greek (i.e., no man is Greek).

However, this kind of rearrangement is impossible in natural language. The general logical form of predication in natural language is '*(np₁, ... npₙ) is P*', where every '*npᵢ*' is a noun phrase which may be quantified. Quantifiers are thus syntactically part of the subjects, and, in contrast to the predicate calculus, sentence negation cannot come *between* a quantifier and the predicate. In order to have negation precede quantification, so to say, natural language has to distinguish syntactically between two ways of predicating: affirmative and negative. This distinction is typically achieved by means of different copulas, affirmative ('is', 'are', etc.) and negative ('isn't', 'is not', 'aren't', etc.); or it may be achieved by indicating affirmative predication by the lack of any copula, and negative predication by means of a negative one (as is partly the case with the present tense in Hebrew); and perhaps some languages achieve this distinction by other means.

A negative copula in natural language is parallel to sentence negation coming between quantification and predication in the predicate calculus. That is why the two predicate calculus sentences above translate, first,

> Some men are not Greek,

where the *copula* is negative; and

> It's not the case that some men are Greek,

where the *sentence* is negated. The copula, redundant from the point of view of the predicate calculus (with either unary or generalized quantifiers), is essential for natural language. We once again see the far-reaching logical and semantic implications of plural reference.

The distinction between affirmative and negative predication makes it essential for us to distinguish the two in our symbolism. A sentence with negative predication will be written as '*(np₁, ... , npₙ) isn't P*'. In case all noun phrases are definite singular ones, there is no distinction between negating the copula and negating the sentence, and this can serve to define the relation between them:

> If every '*npᵢ*' is a definite singular noun phrase, then '*(np₁, ... npₙ) isn't P*' is synonymous with 'It's not the case that *(np₁, ... npₙ) is P*'.

We thus acknowledge two kinds of basic sentences, affirmative and negative, their form being '*(np₁, ... npₙ) is P*' and '*(np₁, ... npₙ) isn't P*', respectively. These sentences are basic in the sense that they are not compounded out of other sentences. Later on we shall consider more complex sentences, the product of combining such basic sentences by means of sentential connectives.

In distinguishing two kinds of non-compounded sentences our logic and semantics follow Aristotle's, not Frege's. Aristotle too maintained that there are two kinds of non-compounded propositions, affirmative and negative (*On Interpretation* 5). In Frege's calculus, by contrast, the basic sentence is always affirmative, negation being limited to sentence negation. Unlike Aristotle, however, we allow not only one-place predicates, but also many-place ones.

Moving now to predicates with several quantified noun phrases among their subjects, we meet the same distinctions again. On the one hand,

Not[every *x*: Man *x*][some *y*: Woman *y*](Loves *x*, *y*)

translates

It's not the case that every man loves some women,

where in both cases the sentence is negated. While on the other hand,

[Every *x*: Man *x*][some *y*: Woman *y*]Not(Loves *x*, *y*)

translates

Every man doesn't love some women,

where the copula is negated – natural language's parallel of sentence negation coming between predication and quantification in the calculus.

The negation of the copula can sometimes be read either as sentence negation or as predication negation, the correct reading often depending on emphasis or intonation. In 'John doesn't love *three* women; he loves *four* women', the negation is sentence negation. The first sentence is synonymous with 'It's not the case that John loves three women'. However, in 'John doesn't love three women: Jane, Penelope, and Mary', the negation is predication negation. From our point of view what is important is not that the negative copula can be used as sentence negation, but that it can be used as predication negation, something unnecessary for the predicate calculus.

The former examples involving two quantified noun phrases raise the following question. In addition to the two possibilities mentioned above, the predicate calculus allows of a combination of negation and quantification of one further kind, where sentence negation comes between two quantifiers. For example:

1 [Every *x*: Man *x*]Not[some *y*: Woman *y*](Loves *x*, *y*).

How can natural language capture this logical possibility? We can use the artificial construction, 'For every man it's not the case that he loves some women', but this looks again too much like Calculish. If we confine ourselves to the use of quantified noun phrases as subjects of predication, either positive or negative, then there is no third possibility in addition to those mentioned above.

This, however, does not indicate any expressive incompleteness of natural language. Sentence (1) is synonymous with the following sentence:

[Every *x*: Man *x*][every *y*: Woman *y*]Not(Loves *x*, *y*),

and this sentence translates the natural language sentence

Every man doesn't love any woman.

In general, since by alternating 'some' and 'every' we can move all sentence negations coming between two quantifiers in the calculus either to the beginning or to the end of the quantifier string, any logical possibility expressible in the calculus with 'some' and 'every' is expressible in natural language as well.

Of course, there are other quantifiers besides 'some' and 'every'. To show that natural language is not lacking in expressive power, we should show that for any two quantifiers 'q_1' and 'q_2', natural language can say what is said by a sentence of the form

$$[q_1 x: x \text{ is } A]\text{Not}[q_2 y: y \text{ is } B]R(x, y).$$

This is generally possible because the formula part 'Not[q x: x is A]' is substitutable by '[$\bar{q}$ x: x is A]', where '$\bar{q}$' is the *complementary* quantifier to 'q'. '$\bar{q}_2$' is complementary to 'q_1' if it includes all quantities precluded by 'q_1'. For instance, 'at most two' is complementary to 'at least three'; 'either less or more than seven' is complementary to 'seven'; 'not many' is complementary to 'many'; 'not all' to 'all'; and 'none' to 'some'.

In this way a sentence of the above form is synonymous with

$$[q_1 x: x \text{ is } A][\bar{q}_2 y: y \text{ is } B]R(x, y),$$

which is directly translatable into natural language. For instance, the sentence '[Some x: x is a man]Not[at least three y: y is a woman]Love(x, y)' is synonymous with '[Some x: x is a man][at most two y: y is a woman]Love(x, y)'; and this latter sentence translates the natural language sentence 'Some men love at most two women'. Similarly, '[Some x: x is a man]Not[seven y: y is a woman]Love(x, y)' is synonymous with '[Some x: x is a man][more or less than seven y: y is a woman]Love(x, y)', which translates the natural language sentence 'Some men love either more or less than seven women'. Lastly, '[Some x: x is a man]Not[some y: y is a woman]Love(x, y)' is synonymous with '[Some x: x is a man][no y: y is a woman]Love(x, y)', which translates the natural language sentence 'Some men love no woman'.

Generally, if sentence negation comes between two quantifiers of the predicate calculus, we can delete it and replace the quantifier following it by its complementary, without affecting meaning. Once all such negations are treated in this way, the resulting sentence can be directly translated into natural language. Accordingly, natural language is not lacking in expressive power in this respect compared with the predicate calculus.

Chapter 8

Pronouns, Variables, and Bound Anaphors

8.1   Pronouns and other Definite Noun Phrases as Alleged Variables

Before proceeding to a comparison of pronouns and variables, I shall note another distinction between the referring expressions of the predicate calculus and those of natural language, a distinction that I shall use shortly. While the latter can have descriptive content, the former lack it. By 'descriptive content' I mean what Strawson (1950, p. 21) characterized as conventional limitation of reference to things of a certain general kind, or possessing certain general characteristics.[1] The rule for the referential use of the expression 'tall man' is that it should be used to refer to tall men, of 'I' that it should be used to refer to the speaker, of 'these children' that it should be used to refer to children, and so on. (Excepting cynical, metaphorical, or other secondary uses, where the reference is to particulars of which the descriptive content of the expression, although still relevant to what is conveyed, is not true.) By contrast, the proper names and variables of the predicate calculus can be used to refer to anything, be it a person, an object, an event or what have you. (Referring expressions with descriptive content can be added to the calculus – the iota operator, for instance, introduces such expressions; but given my purpose in the discussion below, it is important to note that they are not part of the standard version of the calculus.)

Let us now proceed to a comparison of pronouns and variables. It is often claimed that the function variables have in the predicate calculus is occupied in natural language by pronouns. 'Variables are essentially pronouns', asserts Quine (1987, p. 237). Such claims are based on the function pronouns have in some sentences, exemplified, for instance, by the way 'it' functions in

> If a lioness notices a wounded animal, she will try to catch *it*.

However, this function *is not limited to pronouns*. Definite descriptions can function in this way as well. Observe, for example, the way 'the antelope' and 'the zebra' function in

> If a lioness notices a healthy antelope and a wounded zebra, she will
> ignore *the antelope* and try to catch *the zebra*.

---

[1] Strawson originally called this feature descriptive *meaning*, but later changed his terminology to descriptive *content* (1986, p. 92).

Logicians assumed that only pronouns are the variables of natural language because
(i) pronouns have minimal descriptive content, and are therefore similar in this respect
to the calculus' variables, and because (ii) they are the part of speech most frequently
used in the alleged variable role (for reasons discussed below, section 8.3). Assertions
like the one quoted above from Quine, or his claim that 'pronouns are the basic media
of reference' (1948, p. 13), based on that assumption, are therefore mistaken.[2]

8.2   Variables versus Bound Anaphors

Contrary to what is commonly maintained, the semantic function of a pronoun or of
any other noun phrase used in the alleged variable role in a quantified sentence is not
that of a variable. In the predicate calculus variables have a distinct function: they are
part of the quantified construction, and in any substitution instance of the quantified
sentence they are substituted by a constant, a proper name. By contrast, noun phrases
in the alleged variable role do not have any special function in the construction of
quantified sentences. The rules of their functioning resemble those of truth-functional
operators in quantified sentences in the following way: the relation of truth-functional
operators in quantified sentences to truth-functional operators in non-quantified
sentences is like that of noun phrases in the alleged variable role to anaphoric noun
phrases in non-quantified sentences. Let me explain.

Consider the two sentences:

1   If Paul bought this donkey, he vaccinated it.
2   If Paul bought a donkey, he vaccinated it.

At least for the sake of my argument, we can regard the conditional operator in
sentence (1) as truth-functional: sentence (1) is false if and only if its antecedent is
true and its consequent false. However, the conditional operator in sentence (2)
cannot be explained in exactly the same way. This is because the consequent of this
sentence (and perhaps its antecedent as well) does not have a truth-value. Only the
sentence as a whole is either true or false. Yet we do not need to forge any new
explanation for the meaning of the conditional operator in (2). The general
explanation of the relation of the meaning of quantified constructions to non-
quantified ones, together with the explanation of the meaning of conditional
operators in non-quantified sentences, already explain the contribution of the
conditional operator to the meaning of sentence (2). Sentence (2) is true if and only
if every definite singular noun phrase referring to a donkey would yield a true
sentence, if it is substituted for the phrase 'a donkey'. Only the relation of the
meaning of the quantified construction to that of the relevant non-quantified ones

---

[2] The fact that not only pronouns can be used in the alleged variable role was noted in passing
by Evans (1977a, p. 103). Evans, however, made very limited use of this fact, and he seems not
to have noticed its implications for the common philosophical views, demonstrated by my
quotations from Quine. Evans does not even mention this fact in a later paper on pronouns
(1980), although that paper was intended primarily for linguists.

was mentioned in this explanation – we did not even have to mention the existence of a conditional operator in these sentences. Once the substitution has been made, the conditional operator contributes to the meaning of the sentence in the same way that it did in sentence (1). A truth-functional operator "bound" by a quantified noun phrase does not have a meaning distinct from that of the corresponding unbound operator.

Similarly for noun phrases in the alleged variable role, which I shall call *bound anaphors*. Consider again sentences (1) and (2). The pronoun 'it' in sentence (1) refers to the donkey to which 'this donkey' refers. It is an anaphoric noun phrase; i.e., its reference is determined as that of a noun phrase appearing (usually) earlier in the sentence or discourse, its *source*. Now the pronoun 'it' in (2) cannot be explained in exactly the same way; in sentence (2), 'it' does not refer to any donkey, nor to anything else. However, its contribution to the meaning of sentence (2) does not need any additional explanation. The relation of the meaning of 'it' in (2) to 'it' in (1) is that of the conditional operator in (2) to that in (1). The general account of the dependence between the truth-value of a quantified sentence and that of its instances says that (2) is true if and only if all substitutions of definite noun phrases referring to donkeys for 'a donkey' yield true sentences. And in any such substitution, 'it' functions as an anaphoric noun phrase. The rule of use necessary for the understanding of a quantified sentence because of the presence of a bound anaphor in it is the same rule of use needed for an unbound anaphor.[3]

If we consider the relation of a quantified sentence of the predicate calculus to its instances, we see that in each of its instances the variable is substituted in the sentential function following the quantifier by a singular definite noun phrase. For instance, a substitution instance of

$$(\text{Every } x)(\text{Man } x \rightarrow \text{Mortal } x)$$

is

$$\text{Man}(\text{Paul}) \rightarrow \text{Mortal}(\text{Paul}).$$

This is the case for binary and restricted quantification as well. For instance, a substitution instance of

$$[\text{Every } x: \text{Man } x]\text{Mortal } x$$

is

$$\text{Mortal (Paul)}.$$

---

[3] The account supplied here of the relation between bound and unbound anaphors is not new. Its essential idea, along with the comparison to bound and unbound connectives, can be found in Evans' papers of 1977a and 1980 (§ 3). The conclusions I draw below, however, concerning the difference between bound anaphors and variables, is not to be found there, nor, to the best of my knowledge, in any other publication.

By contrast, bound anaphors are not substituted by other definite noun phrases in the analogous cases, as is demonstrated by sentences (1) and (2). Indeed, occasionally bound anaphors need to be substituted, as in the following case:

> Every child loves *his* mother
> Mary loves *her* mother.

But this substitution is for grammatical reasons – in this case a match in the gender of anaphor and source – and not for semantic reasons. It resembles the following grammatical substitution of 'is' by 'am':

> Every man is mortal
> I am mortal

By contrast, substitution in the case of the predicate calculus' variables is part of the meaning of what it is to be a variable.

An anaphoric noun phrase is called 'bound' if it is anaphoric on a quantified noun phrase. Its relation to an unbound anaphoric noun phrase is that of a bound connective to an unbound one. It is not the relation of a bound variable to an unbound one. Bound anaphors are not the variables of natural language. Natural language has no variables.

Since variables take in translation the place of plural referring expressions in natural language, they often appear in translations of sentences in which there is no bound anaphor. The predicate calculus translates a plural referring expression by a predicate, and the variable then occupies both the place occupied in the translated sentence by the plural referring expression and by its bound anaphors, if there are any. Consider, for instance, the following sentence and its translation:

> All men are mortal
> (all x)(Man $x \rightarrow$ Mortal x).

There is no bound anaphor in the first sentence. By contrast, the variable appears twice in the sentential function of the translation. That is because the referring expression 'men' has been substituted by the predicate 'Man'. Consequently, the variable, which occupies the referential places, has to appear once with 'Man' predicated of it, to limit the relevant particulars to men alone; and again with 'Mortal' predicated of it, since it takes over the function of 'men' as that of which mortality has been predicated. This is also true of the appearance of variables in other forms of quantification, e.g., restricted quantification, where the translation is '[All x: Man x]Mortal x'. We again see how different variables are from what is supposed to be their parallel in natural language.

8.3   Rules for the Choice of Anaphors

If we examine the rule that determines which noun phrase should be used as an anaphor (whether bound or not) in a given sentence, we shall understand why pronouns are the noun phrases most commonly used as anaphors. We shall thus understand why they were mistakenly taken to be the bound anaphors of natural language, and consequently its variables. I shall therefore try to formulate this rule in the present section. (As is the case with all or most rules of natural language, the one specified below allows of exceptions for special purposes or in special contexts.) It will be seen that the rule for the choice of anaphors is in accord with Grice's conversational maxims, as formulated in his 'Logic and Conversation' (Grice, 1967, chapter 2).

As a first approximation, we can say that the descriptive content of the anaphoric noun phrase should be either equal to or less specific than that of its source. Consider first the use of the sentence

> If John bought the car, he made a good deal

in which 'he' denotes John. In that use, 'he' is anaphoric on 'John', and its descriptive content is less specific: while 'John' is used to refer to men called 'John', 'he' is used to refer to any male person. The same observation applies to the use of 'the man' and 'the child' in the sentence

> If the man in the green jacket and the child running towards
> him will hug each other, then the man is probably the child's
> father.

On the other hand, the following sentence, if used to make the same statement as the previous one, is defective:

> If the man and the child will hug each other, then the man in
> the green jacket is probably the father of the child running
> towards him.

This sentence is defective because the descriptive content of the anaphoric noun phrases is *more* specific than that of their sources: every man with a green jacket is a man, but not *vice versa*. The anaphoric noun phrase should have the same or a less specific descriptive content.

More accurately, the anaphoric noun phrase should have the *minimal* descriptive content which avoids ambiguity. That is why the sentence:

> If the man in the green jacket standing under the sign is
> waiting for the bus, then *the man in the green jacket* will
> soon be gone

is defective. Ambiguity would be avoided by the minimally specific pronoun 'he':

> If the man in the green jacket standing under the sign is
> waiting for the bus, then *he* will soon be gone.

While the use of 'the man in the green jacket' is appropriate in

If the man in the green jacket standing under the sign and
the man smoking a pipe knew each other, then *the man in
the green jacket* wouldn't have ignored the other one.

Ambiguity would have arisen if the less specific 'he' had been used instead of 'the man in the green jacket'.

Since pronouns usually have the minimal descriptive content that would avoid ambiguity, pronouns are the noun phrases most commonly used as anaphors.

8.4   Conditional Donkey Anaphora

I claimed above that the pronoun 'it' in the sentence

1   If Paul bought a donkey, he vaccinated *it*

is a bound anaphor. This claim is debated in the literature. Beginning with Evans (1977a), some interpret 'it' as a referring expression, designating the donkey that Paul bought. That is, they claim that the consequent presupposes the truth of the antecedent, i.e., that Paul bought a donkey; the pronoun 'it' is then used to refer to that donkey. Following Evans, such a pronoun is called an *E-type* pronoun. By contrast, if 'it' is a bound anaphor, it does not refer to any donkey. Only if 'a donkey' is substituted by an expression that designates a donkey does 'it' refer to a donkey.

Those who deny that 'it' in sentence (1) is a bound anaphor do not maintain that bound anaphors do not exist in natural language. I believe it is generally accepted that, e.g., 'himself' in 'Every man loves himself' *is* a bound anaphor. And my distinctions between bound anaphors and variables would apply to this case as well. Moreover, it is not essential for my claims about the nature of bound anaphors vis-à-vis variables to determine which constructions are bound anaphors and which are referring expressions of other kinds. So it might seem that the debate about the nature of 'it' in sentence (1) should not interest us here. Any other example of a bound anaphor would serve to illustrate the points relevant to my purposes.

However, this debate does have some implications relevant to us here. Many examples support the theory that E-type pronouns, or E-type noun phrases generally, exist in language. Now the view that 'it' in sentence (1) is a bound anaphor has met with a difficulty (specified below) that is by-and-large resolved if 'it' is interpreted as an E-type pronoun. But the application of the E-type theory to this case supported the view that bound anaphors can appear only in a grammatically very limited class of sentences. In particular, it supported the view that anaphors cannot be bound across sentential connectives.

Now this conclusion is relevant to us here, for it seems to limit the expressive power of natural language. If natural language is to have expressive power similar to that of the predicate calculus in all that concerns quantified constructions, then it seems it has to be able to bind anaphors across sentential connectives. So I shall try to show that bound anaphors across sentential connectives do exist in language. Moreover, I shall attempt to support the view that 'it' in sentence (1) is such a bound anaphor. And lastly, I shall try to resolve the difficulty which supported the E-type interpretation of that 'it'.

Before proceeding with this discussion, a note on its limited scope. Anaphora as in sentence (1) is called 'donkey anaphora'. In that sentence the source and the donkey anaphor 'it' are in different sub-sentences, 'Paul bought a donkey' and 'he vaccinated it', respectively, sentences connected by a conditional. A different type of sentence that is also considered as involving donkey anaphora is the following:

Every man who buys a donkey vaccinates it.

Here the source is in the defining clause that is part of the subject term, and the donkey anaphor is the direct object of the verb. By contrast to the anaphora in sentence (1), in this sentence no sub-sentence contains only the source or only the anaphor. However, the two kinds of donkey anaphora are closely related. The latter sentence, for instance, is synonymous with 'If a man buys a donkey he vaccinates it' – where the donkey anaphora is across a conditional. A comprehensive account of donkey anaphora should account for these two kinds of sentence, explaining their interrelations and distinctions. This, however, is not my purpose here. The deductive system in Part III of this book is not meant to apply to sentences in which a noun has a defining clause that itself contains a quantified noun phrase. I shall therefore discuss here only the conditional donkey anaphora.

We can now proceed with our discussion of E-type noun phrases and the conditional donkey anaphora. First, let us examine some E-type noun phrases, in order both to explain the phenomenon more clearly and to demonstrate the existence of E-type noun phrases which are not pronouns. Consider the sentence

2   Some students are intelligent, and they are also nice.

If 'they' were a bound anaphor, then only the sentence as a whole would have a truth-value. However, a situation in which both conjuncts are, for instance, true, can easily be imagined. Moreover, if 'they' were a bound anaphor, then (2) would be true if and only if '*A* is intelligent and he is also nice' were true for some '*A*'s, where each '*A*' designates a student. That is, (2) would be synonymous with

3   Some students are both intelligent and nice.

But that is not what (2) means. A consequence of (2) is that *all* students who are intelligent are nice, while (3) is consistent with some students being intelligent but not nice. So 'they' in (2) is not a bound anaphor.

'They' in (2) is an E-type pronoun. The second conjunct in (2) presupposes the truth of the first conjunct, and it uses 'they' to designate those students which make the first conjunct true, i.e., the students who are intelligent. In case the first conjunct is false, 'they' does not designate anybody.[4]

---

[4] A similar theory of some noun phrases, *relativa grammaticalia* in Scholastic terminology, was developed by Buridan in his *Tractatus de Suppositionibus*; see Hülsen (2000, § 6). I do not

Consider now the sentence

Some students and a few professors are intelligent, and the students are also nice.

Arguments similar to those just mentioned strongly support the view that 'the students' above is an E-type noun phrase. The phenomenon of E-type reference is generally mistakenly ascribed in the literature to pronouns alone, because they are the noun phrases most frequently used in this way, for reasons specified above (§ 8.3).

Secondly, let us see some examples of anaphors bound across sentential connectives. The considerations against considering 'they' in (2) as a bound anaphor do not apply to the following sentences:

Every cat likes milk, unless it is Siamese.
Some children will get the flu only if they are not vaccinated.

For instance, during a debate on whether to vaccinate children against the flu, one who supports vaccination could argue: 'Some children will get the flu anyway, but some children will get the flu only if they are not vaccinated.' In this case, it makes no sense to interpret the pronoun 'they' as an E-type pronoun, referring to the children who verify the antecedent – i.e., all the children who will get the flu. In the context described above it was stated that some children will get the flu even if they are vaccinated; so if 'they' designated *all* the children who will get the flu, it would have been stated, among other things, that the children who will get the flu even if they are vaccinated will get the flu only if they are not vaccinated – but no such inconsistency is involved in the actual statement. By contrast, if we interpret 'it' and 'they' as bound anaphors, the meaning of the sentences is unproblematic.

So there are good reasons for maintaining that there are anaphors bound across sentential connectives. In addition, the interpretation of noun phrases which function like 'it' in 'If Paul bought a donkey, he vaccinated it' as E-type noun phrases is problematic. This is mainly for two reasons, both first noticed by Irene Heim.

Consider first the sentence

If a man is in Athens, he is not in Rhodes.

If we interpreted 'he' in this sentence as an E-type pronoun, then it should designate the man who is in Athens. But there isn't just *one* man in Athens, and the antecedent does not supply any reason for preferring one of the men in Athens over the others. So it seems that 'he' cannot designate any man in Athens, and consequently that it is not an E-type pronoun.[5] By contrast, if 'he' is a bound

---

know whether Buridan applied his theory to sentences like (1), but he did apply it to sentences like (2), of which the theory seems correct.

[5] In reply to this objection, Neale (1990, § 6.3) developed a theory according to which the pronoun 'he' in 'If a man is in Athens, he is not in Rhodes', although syntactically singular, is semantically numberless, designating *all* the men in Athens, be they one or many. I.e., that sentence would be synonymous, according to Neale, with the sentence 'If a man is in Athens,

anaphor, then the sentence is true if and only if any substitution of a name designating a man for 'a man' would yield a true sentence, where 'he' would be anaphoric on that name. For instance, 'If Socrates is in Athens, he is not in Rhodes' should be true. This interpretation yields the right meaning, and it does not have the non-uniqueness problem.

Consider next the sentence

> If a man shares an apartment with another man, he shares the housework with him.

If 'he' were an E-type pronoun, it should designate the man who shares an apartment with another man. But assuming the antecedent to be true, there are at least two such men. And again, the antecedent supplies no reason for preferring one of these as the one designated by 'he'. So it seems that 'he' in that sentence is not an E-type pronoun. However, if it is a bound anaphor, this problem of indistinguishable particulars does not arise. If we substitute a name of a man for 'a man', and 'he' is anaphoric on it, any substitution should be true. For instance, 'If Paul shares an apartment with another man, he shares the housework with him' should be true. Again this interpretation yields the right meaning.

So there are good reasons for not interpreting as E-type those noun phrases that function in conditional sentences as 'it' does in sentence (1), 'If Paul bought a donkey, he vaccinated it', while their interpretation as bound anaphors seems plausible. And as we have seen, this would not be the only case of anaphors bound across sentential connectives. To supply this bound-anaphora interpretation with final support, we should resolve the difficulty it involves.

The difficulty is that while the indefinite article in most constructions is synonymous with 'at least one',[6] if 'it' in sentence (1) is an anaphor bound by 'a man',

---

then the man or men who are in Athens are not in Rhodes'. But this does not seem to fit our understanding of that sentence. Moreover, Barker (1997, § 1.1, pp. 197-200) showed that Neale's theory yields wrong interpretations of sentences like 'If a theory is classical, then if it is inconsistent, it is trivial.'

Some linguists, in their attempt to explain the meaning of similar pronouns, such as 'it' in 'Every farmer who owns a donkey beats it', claimed that they denote the *sum individual* designated by the definite description 'the donkey or donkeys he owns' (see references in Kenazawa, 2001, p. 385). But to introduce such artificial and perhaps incoherent individuals in order to explain these common sentences seems misguided, and this theory definitely disagrees with what we understand these sentences to mean. Kenazewa also finds this theory 'counterintuitive' (p. 400).

[6] There are additional constraints on the use of the indefinite article. For instance, one should use it only if one has in mind a specific particular of which what one says is true. This constraint does not exist if one uses instead the quantifier 'at least one' in an otherwise identical statement. Both statements, however, will have the same truth-value.

My description of this constraint might not be entirely accurate, but we need not determine the meaning of the indefinite article any more accurately for this discussion. It should suffice that statements made with the indefinite article followed by a common noun used referentially are true if and only if they are true for at least one particular of those the noun designates.

then the indefinite article functions there as a *universal* quantifier. The usual translation of (1) into the predicate calculus (although defective for other reasons discussed in this work) illustrates this point. Using unary quantifiers, the standard translation of sentence (1) is:

(Every x)(Donkey $x \rightarrow$ (Paul bought $x \rightarrow$ Paul vaccinated x)).

And if we use restricted quantification, it is:

[Every x: Donkey x](Paul bought $x \rightarrow$ Paul vaccinated x).

The fact that the indefinite article should be interpreted as a universal quantifier in such constructions is not a peculiarity of English. The same phenomenon recurs also in Hebrew, for instance, which is grammatically very different from English. So the universal power of the indefinite article in conditional sentences should be given some general semantic explanation.

I shall now try to supply such an explanation. I shall not do that by pointing out some hitherto unnoticed facts, but by illuminating some familiar ones with a new light.

Consider the conditional sentence:

4   If Paul bought a donkey, then he's happy.

Here there is no noun phrase in the consequent that could be either a bound anaphor or an E-type noun phrase. The scope of the noun phrase 'a donkey' is therefore naturally taken to be the antecedent alone; that is, in this case, the minimal sentence containing the noun phrase 'a donkey' that has a truth-value is the antecedent, 'Paul bought a donkey'. And there the indefinite article has its usual meaning, i.e., at least one. Any other quantifier substituted for the indefinite article will also have its ordinary meaning. Using restricted quantification, the translation of (4) into the predicate calculus is:

5   {[An x: Donkey x](Paul bought x)} $\rightarrow$ (Paul is happy).

But suppose we wanted to interpret the noun phrase 'a donkey' in sentence (4) as having wide scope, i.e., as if only the implication as a whole has a truth-value: how should we interpret the indefinite article then? That is, how many substitution instances of sentence (4), formed by substituting singular terms referring to different donkeys for 'a donkey', should be true, if and only if sentence (4) is true? The answer is, *every* substitution. Let us prove that.

Suppose the indefinite noun phrase 'a donkey' in sentence (4), 'If Paul bought a donkey, then he's happy', is interpreted as having narrow scope and as meaning *at least one*. When is sentence (4), thus read, false? Sentence (4) is then false if its antecedent is true and its consequent false. So sentence (4) on this interpretation is false if 'Paul bought a donkey' is true but 'he's happy' is false.

Next, suppose the indefinite noun phrase 'a donkey' in sentence (4) is interpreted as having wide scope and as meaning *every*. When is sentence (4) false on *this* interpretation? It is false if for *some* substitution of a name of a donkey for 'a donkey', the resulting sentence is false. But the resulting sentence is an implication without any quantified noun phrase, and it is therefore false if its antecedent is true and its consequent false. So sentence (4) is false is for *some* such substitution, 'Paul bought a donkey' is true and 'he's happy' is false. But that means that sentence (4), on this interpretation, is false if 'Paul bought a donkey', with the indefinite article meaning *at least one*, is true, but 'he's happy' is false. And this was the case in which sentence (4) on the former interpretation was false.

We thus see that on the two interpretations sentence (4) has the same truth conditions. Both interpretations entail each other, so they are equivalent.

This equivalence is reflected in the predicate calculus by the well-known equivalence between sentence (5) and the following sentence:

[Every x: Donkey x]{(Paul bought x) → (Paul is happy)}.

So the ordinary interpretation of the indefinite article entails its interpretation as a universal quantifier when it appears in the antecedent of conditionals and its noun phrase does not bind anaphors in the consequent. Accordingly, if it is used in conditionals with the same meaning as in sentence (4), and its noun phrase *does* bind anaphors in the consequent – i.e., if it has wide scope – it functions as a universal quantifier. We have proved that the usual interpretation of the indefinite article entails its interpretation as a universal quantifier in conditional donkey anaphora sentences.

So we have explained why the indefinite article, or any other quantifier meaning 'at least one', when used with noun phrases having wide scope in conditional sentences, has the power of a universal quantifier. This is why all the following sentences are read as universally quantified:

If Paul bought a donkey, he vaccinated it.
If Paul bought any donkey, he vaccinated it.
If Paul bought at least one donkey, he vaccinated it.
If Paul bought one or more donkeys, he vaccinated them.

The difficulty on the interpretation of the indefinite article as a universal one in conditional sentences is therefore resolved.

Our considerations are supported by the fact that their generalization to conditional donkey anaphora with other quantifiers is straightforward. Consider, for instance, the sentence

6   If Paul bought two donkeys, he vaccinated them.

If we read the pronoun 'them' in sentence (6) as a bound anaphor, 'two donkeys' should then be taken as having wide scope, and 'two' should mean *any two*. Sentence (6) is then synonymous with the following sentence:

For any two donkeys, if Paul bought both of them, he vaccinated both of them.

This meaning of 'two' can be derived from the equivalence of the two readings of the following sentence:

If Paul bought two donkeys, then he's happy.

The reading of 'two donkeys' as having narrow scope and 'two' as synonymous with 'at least two' is equivalent to the reading of 'two donkeys' as having wide scope and 'two' as synonymous with 'any two'. This equivalence is again reflected in the following equivalence of the predicate calculus:

$(\exists x)(\exists y)$(Donkey x & Donkey y & $x{\neq}y$ & Paul bought x & Paul bought y)$\rightarrow$Paul is happy

$(\forall x)(\forall y)\{$(Donkey x & Donkey y & $x{\neq}y$ & Paul bought x & Paul bought y)$\rightarrow$Paul is happy$\}$

Analogous equivalence exists for any quantifier 'n', and consequently any such quantifier can be interpreted in conditional donkey sentences as synonymous with 'any n' with its noun phrase binding anaphors across the conditional. And this generalization applies to other, non-numerical quantifiers as well; for instance, the sentence 'If Paul bought several donkeys, he vaccinated them' means that for *any* several donkeys, if Paul bought *all* of them, he vaccinated them all.[7]

To properly understand the relation between the quantified noun phrase with its wide scope reading and its bound anaphor in such donkey sentences we have to turn to our discussion of collective predication in the context of quantified noun phrases (§ 7.3). I said there that the substitution rule for collective predication has a more fundamental semantic status than that for distributive predication, since it applies to distributive predication as well, and the specific rule for the latter can be derived from it. We will now see that in the case of donkey anaphora the specific substitution rule for distributive predication cannot be derived from the general substitution rule and that it is invalid.

In cases that do not involve anaphora the specific distributive substitution rule is derived from the general collective rule as follows. 'Two men came to work' is true if 'John and Alison came to work' is; but the latter is true if and only if 'John came to work' is true and 'Alison came to work' is.

---

[7] Stephen Neale failed to notice that the behavior of the indefinite article in donkey sentences is parallel to that of other numerical quantifiers. He considers (Neale, 1990, pp. 225-6) the sentences (1) 'Every man who bought two or more donkeys vaccinated them' and (2) 'If John buys several donkeys he vaccinates them'. In both there is universalization; e.g., every man who bought two or more donkeys vaccinated *every* donkey he bought. But, claims Neale, 'we cannot capture this fact by treating "two or more donkeys" as a wide-scope quantifier – universal or otherwise – that binds "them". If the quantifier were universal, (1) would be equivalent to ["Every man who bought a donkey vaccinated it"], which it is not.' But I believe Neale is mistaken on this point: a universal *plural* quantifier can capture the meaning of (1), as the following sentence, synonymous with (1), demonstrates: 'For any two or more donkeys, every man who bought *all of them* vaccinated them.' And I have also just demonstrated this fact in the text for *conditional* donkey sentences with the quantifiers 'two' and 'several'. Thus, *pace* Neale, all this suggests that in order to understand donkey anaphora we *don't* 'need to think about the semantics of the anaphoric [noun phrases]', but about 'the semantics of their antecedents.'

Consider, however, sentence (6), 'If Paul bought two donkeys, he vaccinated them.' We saw that 'two' in sentence (6) means *any* two. Substitute, then, a conjunction of two names of donkeys for 'two donkeys', as our general substitution rule instructs us to do. For instance:

7   If Paul bought Platero and Pompa he vaccinated them.

We can deduce from this sentence the following one:

> If Paul bought Platero and he bought Pompa, then he vaccinated Platero and he vaccinated Pompa.

But we cannot deduce sentence (8) from sentence (7):

8   If Paul bought Platero he vaccinated Platero, and if he bought Pompa he vaccinated Pompa.

Suppose Paul bought Platero, did not buy Pompa, and did not vaccinate Platero. Then sentence (8) is false while sentence (7) is not false. However, sentence (8) is the conjunction of the two sentences that we would have formed had the specific distributive substitution rule been valid in such cases. So in the case of donkey anaphora one should always apply the general, collective substitution rule.

Our collective substitution rule, in contrast to the distributive one, also yields the correct interpretation of the famous example, 'proposed by Geach and perfected by David Kaplan, in private correspondence' with Quine (Quine, 1982, p. 293): 'Some people admire only one another'. This sentence is true if and only if some definite noun phrase, designating several people, yields a true sentence if substituted for 'some people'.

The fact that in some cases the collective substitution rule yields the right interpretation, while the distributive one does not, will not affect the deductive system developed in Part III of this book, despite the fact that this system is based on the *distributive* substitution rule. This system applies only to the universal and particular quantifiers, with *singular* bound anaphors. In the context of conditional donkey anaphora we shall apply it only to the indefinite article or to the quantifier 'any'; and these were shown in this section to function in that context as universal quantifiers interpreted according to the distributive substitution rule.

8.5   Predicate Connectives, and Bound Anaphora across Sentential Connectives

I shall add a few remarks on other uses of anaphors bound across sentential connectives. Firstly, we saw that many quantifiers of the form '*q*' mean *any q* in the conditional donkey anaphora. In particular, 'some' means *any some*, as in the sentence 'If Paul bought some donkeys, he vaccinated them'. However, it seems that in order to have sufficient expressive power, natural language has to bind anaphors across conditionals by means of the quantifier 'some' as well, meaning *some* and not *any some*.

This is done by using, instead of the ordinary conditional, its transposed form; that is, by saying, instead of 'If p then q', 'q if p'. For instance, 'some' in the following sentence means *any some*:

> If some children are not vaccinated, they will get the flu.

The sentence says that *all* the children who are not vaccinated will get the flu. But in the next sentence, 'some' indeed means *some*:

> Some children will get the flu if they are not vaccinated.

This sentence allows for the case in which some children will not get the flu even if they are not vaccinated, and for the case in which some children will get the flu even if they are vaccinated. The sentence is true in case it is true for some substitutions of expressions designating individual children for 'some children' (with 'they are' changed to fit number and gender). I.e., some sentences like, for instance, 'Mary will get the flu if she is not vaccinated', should be true.

Next, it should be noted that anaphors usually cannot be bound across the conjunction in conjunctive sentences. The pronoun in the following example is an E-type pronoun:

> Some students are nice, and they are also intelligent.

'They' in this sentence designates all nice students. To say of some students that they are nice and intelligent, while leaving as a possibility that some nice students are not intelligent, we have to use predicate conjunction:

> Some students are nice and intelligent.

In order to have full expressive power, natural language has to be capable of using E-type noun phrases across conjunction. It also has to have the parallel of the predicate calculus quantification over conjunctive sentential functions. And it also has to avoid ambiguity. Accordingly, since predicate conjunction is parallel to that structure of the predicate calculus, noun phrases in conjunctive sentences of the above construction are usually E-type.

The case with disjunction seems to be more complex. Consider the following sentence:

1   Some students will be late, or they may not arrive at all.

'They' in sentence (1) cannot designate the students who will be late, because then it will not be possible that they may not arrive at all. So 'they' cannot be a simple E-type pronoun – i.e., it cannot designate the students who verify the antecedent sentence. But is it a bound anaphor? If it is, then the sentence would be true if, for instance, there were two students, John and Mary, such that John was late and Mary did not arrive at all. I am not sure that sentence (1) is true in such a case.

On the other hand, one may utter sentence (1) having certain students in mind, for instance, the students who one knows have not arrived yet; and then 'they' in the consequent is used to designate all those students. Such a use would be similar to the use of 'he' in the following exchange (Strawson's example): 'A man has just jumped from the roof!'—'He didn't jump, he was pushed.' Here 'he' is not intended to designate the man who jumped from the roof, as it is assumed there was no such man; rather, it is intended to designate the man whom the first speaker *thought* verified his utterance. If 'they' in sentence (1) is used in this way, would sentence (1) be true in case some of the students who have not yet arrived were late, while others did not come at all?

Luckily, we do not have to decide these questions. It is sufficient for our purpose to note that natural language uses predicate disjunction to achieve the meaning sentence (1) would have were 'they' a bound anaphor:

Some students will either be late or may not arrive at all.

It seems this sentence is true in the case described above, namely, if John is late and Mary does not arrive at all.

Because of the importance of predicate connectives for the expressive power of natural language, a fully-developed deductive system for natural language should consider their role in inferences. I supply some treatment of this topic in section0.

8.6 The Relation between the Truth-Value of a Quantified Sentence and those of Its Instances

I shall conclude this part of the book with a formal definition of the relation between the truth-value of a quantified sentence, which may contain bound anaphors, and the truth-values of its substitution instances. This definition will be of much use in the next part of this book, where a deductive system for natural language is developed. This definition will also generalize the explanation of iterative quantification given in section 6.2.

Suppose we are given a propositional combination C of sentences of the form '$(np_1, \ldots np_n)$ is P' or '$(np_1, \ldots np_n)$ isn't P', $n \geq 1$, where for every i, 'np_i' is a noun phrase, and P is an n-place predicate. The number n need not be the same for all such sentences in C. Suppose the first quantified noun phrase in C is 'np_i', which is of the form '$q\,A$', where 'q' is a quantifier and 'A' a plural referring expression. Suppose further that C does not contain any sentence which contains both 'np_i' and all the bound anaphors of any quantified noun phrase appearing in it. Then C is true if and only if, if we substitute 'c' for 'np_i', where 'c' is a definite singular noun phrase referring to a particular to which 'A' refers, and there is exactly one 'c' which refers to any particular to which 'A' refers, and any bound anaphor of 'np_i' is now anaphoric on 'c', we get a true formula for q substitutions.

As explained in section 7.3, this rule applies to cases where the predication is distributive and the quantification iterative. The rule is applicable in cases where the iteration order is the order in which the noun phrases are written or uttered; in cases where the iteration order is different (see section 7.2), the rule should be applied according to that different order.

The clause saying that C should not contain any sentence which contains both 'np_i' and all the bound anaphors of any quantified noun phrase appearing in it is intended for cases like the following. If C is 'It is not the case that all men know English', then for it to be true it need not yield a true sentence for all substitutions of 'all men' by proper names of men. On the other hand, if C is 'All men know English', then if it is to be true it *should* yield true sentences for all substitutions of proper names of men for 'all men'.

Notice that C can be contained in another sentence – in that case C's truth-value will be determined, relative to those of it's substitution instances, independently of the structure of the sentence in which it is contained. The truth-value of the containing sentence can then be determined relying on C's truth-value. For instance, in order to determine the truth-value of 'It is not the case that all men know English', we first determine that of 'All men know English', which complies with the conditions specified in the rule. The truth-value of the former sentence, being a negation of the latter, can then be determined.

Our rule mentions all the bound anaphors of *any* quantified noun phrase appearing in the sentence, not only bound anaphors of the *first* quantified noun phrase appearing in it, because of cases like the following. Consider the sentence 'If a butcher buys a donkey, the donkey is doomed'. The first quantified noun phrase in this sentence is 'a butcher', which has no anaphor. The minimal sentence which contains it and all its bound anaphors is therefore the antecedent, 'A butcher buys a donkey'. But the antecedent, as contained in the above conditional, does not have a truth-value of its own – only the conditional as a whole has a truth-value. This is because the noun phrase 'a donkey', appearing in the antecedent, has a bound anaphor 'the donkey' – in the consequent Our rule is appropriate for such sentences.

The above rule is a semantic rule, which will be used in the proof of the consistency of the deductive system developed in the next part of this book. It shares this function with Tarskian definitions of truth in a model. However, in contrast to Tarskian semantics, I did not specify in my rule the *conditions* in which sentences are true. What I did was to specify the *relations* between the truth-*values* of a quantified sentence and those of its instances. Validity is a question of the possible relations between the truth-values of the premises and those of the conclusion. A semantic rule tailored to questions of validity can therefore have more modest aims than those of Tarskian semantics.

My formalization departs from that of the predicate calculus in some obvious ways. I use a copula, both positive and negative; concepts may appear as parts of subjects and as predicates; quantifiers are parts of noun phrases and do not operate on sentential functions; and I use no variables. But I also depart from the common formalization of Aristotelian logic. That logic formalized the four sentence types it acknowledged, universal and particular affirmation and universal and particular negation, as '*SaP*', '*SiP*', '*SeP*' and '*SoP*', respectively. That is, the quantifier and the copula were fused in its formulas. Universal quantification cum affirmation, for instance, were expressed by the symbol '*a*': '*SaP*'. By contrast, in my formalization quantifiers are parts of noun phrases, which are subjects, while the copula expresses the mode of predication. Aristotelian logic's '*SaP*' becomes 'Every *S* is *P*' in my formalization. In this I am closer to natural language than Aristotelian logic was.

Towards the end of Chapter 5 I discussed semantically derived nouns like 'wisdom' and 'philosophy'. I argued there that it is hardly justified to consider their uses in sentences like 'Wisdom is a virtue' referential. I also added that nevertheless, their incorporation in the same place in the *syntactic* framework is conditioned by the applicability of parallel syntactic transformations and derivations to them. For instance, from 'Wisdom is a virtue' and 'Every virtue is rare' we can infer 'Wisdom is rare', by the same derivation principles we use in deriving 'Peter is mortal' from 'Peter is a man' and 'Every man is mortal'. However, my above formulation of the relation between the truth-value of a quantified sentence and the truth-values of its substitution instances explicitly mentioned *referring* expressions. To make it applicable to sentences in which the subjects are semantically derived nouns we should modify the relevant parts of that formulation as follows:

Suppose the first quantified noun phrase in C is 'np_i', which is of the form '$q\,A$', where 'q' is a quantifier and 'A' a semantically derived noun. Then C is true if and only if, if we substitute 'c' for 'np_i', where 'c is an A' is true and every A is named by exactly one 'c', we get a true formula for q substitutions.

The clause 'every A is named by exactly one "c"' in this formulation is obviously a semantic compromise. What does it mean, that an expression *names* an A? We cannot say that the expression *refers* to an A, since I maintained that the use of 'refer' here is unjustified. On the other hand, the adequacy of the above formulation requires that every A will have one and only one *name* – but what should 'name' mean here? Indeed, the uniqueness condition can be satisfied while avoiding the problem that now confronts us by demanding that there be no 'c_1' and 'c_2' such that 'c_1 is c_2' is true; but I cannot see how the existence condition can be satisfied while avoiding that problem. So the apparent semantic relation of naming mentioned in the above formulation seems to be both necessary and devoid of content.

However, perhaps the merely apparent content of 'name' in my formulation is not a shortcoming. Semantically derived nouns are semantically heterogeneous. 'Wisdom', 'philosophy' and 'the fall of Constantinople' may have nothing semantic in common. The relation of 'wisdom' to wisdom is different from that of 'the fall of Constantinople' to the fall of Constantinople; and to say that both are names, designators or expressions used to mention something, disguises their diversity rather than uncovers any uniformity. Perhaps what can be said is that semantically derived nouns can be incorporated in quantified constructions only if a semantic relation can be substituted for that of reference in our original formulation, in a way that preserves the validity of our formulation. Our understanding of quantified

Logic & Natural Language

constructions with semantically derived nouns will then depend on our ability to make the required substitutions. My formulation is actually a *schema*, to be filled with different semantic content for different semantically derived nouns, and not a description relying on "the" semantic relation characterizing such nouns.

PART III

A DEDUCTIVE SYSTEM FOR NATURAL LANGUAGE

Chapter 9

Derivation Rules and Consistency

9.1 Some General Considerations

In this part I develop a deductive system for natural language, which is meant to be comparable in its deductive power to first order predicate calculus. I shall develop the system for only some of the sentences of natural language, and for only some of the logical relations between these sentences. The sentences to which my system will apply are propositional combinations of predications. The predicates will have as subjects either definite singular noun phrases or quantified noun phrases.

The quantifiers I shall discuss are 'some' and 'every'. Although 'every' and 'all' are synonymous in many of their uses, I prefer to use 'every', since it rules out collective readings. 'All the children lifted the table' may mean either that each child lifted the table by himself (distributive predication), or that they lifted it together (collective predication) – where it is doubtful whether it is correct to say of each child that he or she lifted the table. By contrast, in 'Every child lifted the table' the predication is explicitly made with reference to every child.

The sentences to which my system will apply will thus include, for instance:

> John is tall. Some men are tall. John loves Mary. Every man loves Mary. Every man loves some women. Every man gives some presents to some women. Every cat likes milk, unless it is Siamese

As demonstrated in Chapter 8, in some propositional combinations some modes of generalization are expressed with 'a' or 'any', and not with 'every' or 'all'. Accordingly, sentences like the following will be discussed as well:

> If a man buys a donkey, he vaccinates it.
> If any man loves any other man, then that other man loves the first one too.

There are different modes of predication. We can say that John *is* clever, that John *was* clever, that he *may be* clever, and so on. These different modes of predication create various problems for a deductive system. For instance, the valid syllogism:

> All philosophers are Greek. All Greeks are intelligent. Therefore, all philosophers are intelligent,

becomes invalid if we substitute past tense for present tense:

All philosophers were Greeks. All Greeks were intelligent. Therefore, all philosophers were intelligent.

That is because the Greeks that the philosophers were may not be those contemporary Greeks who were intelligent. Like the standard version of first order predicate calculus, I shall not address temporal and modal issues in my system. The predication considered in my system is, as a rule, in present tense. I have applied my system to sentences containing past or future predication only in a few cases, which do not raise any special issue relating to tense.

A significant difference between my system and the predicate calculus is that I shall not supply rules for the well-formed sentences of my system. This is because the language I use is natural, not artificial. The language of the predicate calculus is an artificial language, formed according to explicit rules, and consequently we know the rules which determine which sentences belong to the language. By contrast, for natural languages we know which sentences are correct without learning abstractly formulated rules sufficient to determine correctness. Such a formulation is a notoriously difficult task, which linguists still have to meet.

I shall therefore rely on our knowledge and understanding of English in determining, firstly, which sentences are grammatically correct; secondly, which sentences are propositional combinations of predications of the kind to be discussed; and lastly, in transforming sentences from active to passive, from singular to plural, etc.

This reliance on our knowledge and understanding of natural language is also presupposed when we *apply* the predicate calculus to inferences we actually make. Since these inferences are formulated in natural language, we need, when we try to translate them into the calculus, the same kind of knowledge I assume here in the development of my system. Accordingly, although from a formal point of view the reliance on such knowledge is a disadvantage of my system, it is not a practical disadvantage.

Lastly, although the examples of the applications of my system below are in *English* sentences, the system is meant to be applicable to *any* language whose sentences are propositional combinations of predications with singular referring expressions and quantified plural referring expressions as subjects, sentences whose meaning is in accordance with the rules discussed in the previous parts of this work. Since my purpose in this work is to analyze the semantics and logic of any such language, I did not attempt below to give transformation rules specific to English. For instance, in English, when we deduce 'Some men are mortal' from 'John is mortal' and 'John is a man', we substitute the plural copula 'are' for the singular one 'is'. This is not the case with some other languages in which singular and plural reference, quantification and predication function as in English. In Hebrew, for instance, a copula need not be used at all. On the other hand, in Hebrew the adjective in the corresponding inference is changed from singular to plural form. These are mere syntactic features, language specific, insignificant from a logical point of view. I ignore all such features below.

The fact that I ignore these language specific features makes it impossible to apply my derivation rules *mechanically* to any specific language. For that purpose we should have rules for substituting 'is' by 'are', etc. However, the inferences are carried out *formally*, i.e., we need not know what the predicates involved mean or what the referring expressions designate.

9.2   Basic Characteristics of the System

My deductive system is a system of natural deduction for natural language, based on ideas originally developed by Gentzen (1934-5). My method of writing arguments is based on standard methods such as those found in Lemmon (1965) and Newton-Smith (1985).

I shall write arguments as follows. On the left-most I shall write the line numbers of the sentences on which the present sentence relies. I shall then write in round parentheses the number of the present line. Then the sentence will be written. And lastly, at the right-most, the justification of the present sentence. The line number is also the number of the sentence written at that line.

As will be seen from the derivation rules below, a sentence can rely only on premises. An argument is a *proof* of its last sentence, relying on the premises – if any – on which that last sentence relies. An argument is *valid* if, necessarily, in case all the premises on which a sentence relies are true, that sentence is also true. A calculus is *consistent* if any argument answering its rules is valid. I shall prove that all the rules for writing down new sentences of the calculus developed here preserve validity, and that this calculus is therefore consistent.

I have not proved the *completeness* of my system; i.e., I have not proved the provability in my system of any formally valid argument whose premises and conclusion are sentences of the kind to which my system applies. To partly compensate for that, and to show my system's power, I give below a wide variety of examples of proofs.

I shall occasionally write schemes of arguments, using variables instead of definite singular noun phrases and general nouns. That is, I shall write 'Every *A* is a *B*' instead of 'Every philosopher is a man'. Small italicized letters – *a*, *b*, *c*, etc. – will be used as variables replaceable by definite singular noun phrases, and italicized capital letters – *A*, *B*, *C*, ... – as variables replaceable by predicates.

I now specify the rules for the introduction of premises, for the use of theorems of the propositional calculus, and for the use of other derivation rules of natural language that do not involve quantification.

Firstly, premises. In any line a sentence can be written as a premise. The rule for premises is as follows:

> *Premise.* In any line (i) any sentence can be written, relying on the same line, and justified as a premise. The justification is written 'Premise'.

For instance:

1   (1) John loves Mary   Premise

Since a premise relies on itself, *Premise* clearly preserves validity: if the sentence on which a premise relies is true, i.e., if the premise is true, then the premise is true. Moreover, we shall see that all other derivation rules, i.e., all other rules for writing down new sentences in an argument, always presuppose the existence of previous lines (excluding Identity Introduction; see section 11.5). Thus, the first sentence in any argument is always a premise, and a one-line argument is therefore always valid. The first step in the inductive proof of consistency is accordingly established.

Secondly, derivation rules of the propositional calculus. If a certain sentence is derivable from other sentences by derivation rules of the propositional calculus alone, I shall write the sentence and quote as justification the propositional calculus derivation rule and the numbers of the lines of the sentences from which the sentence is derived. Beside the derivation rule used I shall write '(PCDR)', for *Propositional Calculus Derivation Rule*. The numbers of the sentences on which the derived sentence relies will be written according to the rules of the propositional calculus.

Thirdly, unbound anaphora. If in any sentence a definite noun phrase is anaphoric on a singular referring expression, that anaphor can be substituted by the singular referring expression. For instance, one can infer 'John likes John' from 'John likes himself'. This rule allows the *elimination* of anaphors. And anaphors can also be *introduced*: if a singular referring expression appears several times in a sentence, any but its first appearance can be substituted by a noun anaphoric on the singular referring expression's first appearance. For instance, one can infer 'John likes himself' from 'John likes John'. In both cases the inferred sentence relies on the same premises as those on which the sentence from which it is inferred relies. These rules clearly preserve validity, since the anaphor designates the same particular designated by the noun of which it is an anaphor. Singular referring expressions having the same reference are interchangeable *salva veritate*. To avoid prolixity, I shall occasionally make these transformation without noting them explicitly. I also ignore the cases in which the anaphor precedes its source: we are primarily interested in anaphors bound by quantified noun phrases, and sentences in which such anaphors precede their sources are at least exceptional.

Lastly, as was explained above (p. 91), the negative copula is logically essential for a language in which quantified noun phrases are syntactically among the subjects of a given predicate. This makes the relation of copula negation to sentence negation essential for any logical system of such a language. I shall thus formulate here the rule for substitution of negative copula for sentence negation and *vice versa*. I call this rule 'CN/SN', for *Copula Negation – Sentence Negation*.

> *CN/SN.* If sentence (i) is or contains the sentence 'It's not the case that *(np₁, ... npₙ) is P*', where every '*npᵢ*' is a definite singular noun phrase. Then in any following line (j) the sentence identical to sentence (i), but with '*(np₁, ... npₙ) isn't P*' substituted for 'It's not the case that *(np₁, ... npₙ) is P*', can be written. And if sentence (i) is or contains the sentence

'*(np₁, ... npₙ) isn't P*', where every '*npᵢ*' is a definite singular noun phrase, then in any following line (j) the sentence identical to sentence (i), but with 'It's not the case that *(np₁, ... npₙ) is P*' substituted for '*(np₁, ... npₙ) isn't P*', can be written. In both cases sentence (j) relies on the same premises as sentence (i), and its justification is written 'CN/SN, i'.

(I shall occasionally write 'Not' instead of 'It's not the case that'.) Since 'Not *(np₁, ... npₙ) is P*' and '*(np₁, ... npₙ) isn't P*' are synonymous in case every '*npᵢ*' is a definite singular noun phrase, CN/SN preserves validity.

I now proceed to develop my system.

9.3 Transposition

A relation between different particulars can be represented linguistically in various forms. For instance, the sentences 'John kissed Mary' and 'Mary was kissed by John' represent the same relation. So do the sentences 'John gave this book to Mary', 'This book was given by John to Mary', 'To Mary was this book given by John', etc. This is also the case for 'John is taller than Mary' and 'Mary is shorter than John', and for 'John isn't taller than Mary' and 'Mary isn't shorter than John'. I shall call such variations *transpositions*. They clearly preserve truth, since they are synonymous. Thus, if in a given line in an argument a relation between particulars is represented in some way, it can be transposed in any following line, relying on the same premises. I shall call this derivation rule *Transposition*:

> *Transposition*. If sentence (i) is or contains the sentence '*(np₁, ... npₙ) is/isn't P*', where every '*npᵢ*' is a definite singular noun phrase, then in any following line (j) the same sentence with any transposition of *P* can be written. Sentence (j) relies on the same premises as sentence (i). Its justification is written 'T, i'.

An instance of a simple use of *Transposition*:

1	(1) John loves Mary	Premise
1	(2) Mary is loved by John	T, 1

Transposition preserves validity. Assume that if the sentences on which sentence (i) relies are true, so is sentence (i). Since the transposed part in line (j) means the same as its parallel in line (i), sentence (j) is also true in such a case. But sentence (j) relies on the same premises as sentence (i). So if the sentences on which sentence (j) relies are true, so is sentence (j).

9.4   Universal Elimination

If something is true of every *A*, and *a* is an *A*, then it is true of *a* too. Here *A* stands for any general noun, and *a* stands for any definite singular noun phrase. A more precise formulation of this rule is:

> *Universal Elimination.* Suppose the first quantified noun phrase in sentence (i) is 'every *A*', and sentence (i) does not contain any sentence which contains 'every *A*' and all the bound anaphors of any quantified noun phrase appearing in it. Suppose further that sentence (j) is '*a* is an *A*'. Then, in any line (k), one can write the sentence identical to sentence (i) apart from the fact that in it *a* has been substituted for 'every *A*'. Line (k) relies on the lines on which lines (i) and (j) rely. Its justification is written 'UE, i, j'.

An instance of the application of UE:

1	(1) Every man is mortal	Premise
2	(2) Socrates is a man	Premise
1,2	(3) Socrates is mortal	UE, 1,2

Another example:

1	(1) If a man owns a donkey, he vaccinates it	Premise
2	(2) Paul is a man	Premise
3	(3) Platero is a donkey	Premise
4	(4) Paul owns Platero	Premise
1,2	(5) If Paul owns a donkey, he vaccinates it	UE, 1,2
1,2,3	(6) If Paul owns Platero, he vaccinates it	UE, 5,3
1,2,3,4	(7) Paul vaccinates Platero	MP (PCDR), 6,4

('MP' abbreviates 'Modus Ponens'.) In the passage from (6) to (7) I have also substituted proper names for anaphors. As I said above, to avoid prolixity I shall usually not note such transformations.

The constraint on UE, that sentence (i) should not contain any sentence which contains 'every *A*' and all the bound anaphors of any quantified noun phrase appearing in it, is meant to rule out possibilities such as the following. Given the sentences 'It's not the case that every man is a philosopher' and 'Plato is a man', we should not be allowed to replace 'every man' with 'Plato'. Now the first sentence is the negation of 'Every man is a philosopher', i.e., it contains a sentence which contains 'every man'. Therefore the derivation is barred. Another example: from 'If every man will die within a year the human race will go extinct' and 'George is a man' one should not and cannot infer that 'If George will die within a year the human race will go extinct'.

The constraint on UE is the parallel of the demand in the predicate calculus that the whole sentence be in the scope of the quantifier.

Validity Preservation. We prove that if to a valid argument a sentence is added by UE, the new argument is still valid. Suppose the sentences on which sentence

(k) relies are true. These include the sentences on which both sentences (i) and (j) rely. So sentence (i) is true, and so is sentence (j), '*a* is an *A*'. So according to the substitution rule given above (§ 0, p. 109), we can substitute 'every *A*' by '*a*' and get a true sentence. But this is exactly sentence (k), which is therefore true. So UE preserves validity.

9.5   Universal Introduction

If something is proved for a certain particular, relying only on the fact that it is an *A*, then what is proved is true for every *A*.

> *Universal Introduction.* Suppose sentence (i) is the premise '*a* is an *A*'. Suppose further that sentence (j) contains a single appearance of '*a*', and does not rely on any premise which contains *a* apart from (i). Suppose further that '*a*' in sentence (j) is not preceded by any quantified noun phrase, and that sentence (j) does not contain any sentence which contains '*a*', its anaphors and all the bound anaphors of any quantified noun phrase appearing in it. Then in any following line (k) one can write the sentence identical to (j) apart from the fact that in it 'every *A*' has been substituted for '*a*'. (k) relies on all the premises on which (j) relies, apart from (i). Its justification is written 'UI, j, i'.

An example:

1	(1)	Mary is a woman	Premise
	(2)	Every woman is a woman	UI, 1,1

Since (2) does not rely on any premise, it is a theorem. Similarly, every sentence of the form 'Every *A* is an *A*' is a theorem.

Another example:

1	(1) Every man resembles every man	Premise
2	(2) John is a man	Premise
1,2	(3) John resembles every man	UE, 1,2
1,2	(4) John resembles John	UE, 3,2
1,2	(5) John resembles himself	Anaphora Introduction, 4
1	(6) Every man resembles himself	UI, 5,2

Notice that we cannot infer from (4) that every man resembles John, or that John resembles every man, since 'John' appears more than once in (4). This fact is central for the validity of UI, as can be seen in the proof below (where it is claimed that (j) results from (k) by substitution).

Validity Preservation. Suppose all the premises on which sentence (k) relies are true; we should prove that sentence (k) is true. According to the substitution rule given above (§ 0, p. 109), this is the case if for any definite singular noun phrase '*c*', such that '*c* is an *A*' is true, if we substitute 'every *A*' in sentence (k) by '*c*' we get a true sentence. Suppose, for some given 'c', 'c is an A' is true. Let us substitute any appearance of 'a' in lines (1) to (k-1) in the argument by this 'c'. Since all derivation rules rely only on sameness of definite singular noun phrases and not on the specific definite noun phrase used, the argument generated after this substitution is still valid up to its (k-1) line. Now after the substitution sentence (i) is 'c is an A', which is true according to our assumption. Moreover, all sentences apart from sentence (i) on which sentence (j) relies did not contain 'a', and are therefore true after this substitution. So all sentences on which sentence (j) relies are true, and it is true too. But sentence (j) is the result of substituting 'every A' in sentence (k) by 'c', and (k) is therefore true as a result of such a substitution. And this is the case for any 'c' for which 'c is an A' is true. So sentence (k) is true and UI preserves validity.

9.6 Particular Introduction

If John is tall, and he is a man, then some men are tall:

> *Particular Introduction.* Suppose sentence (i) contains the singular definite noun phrase 'a', which is not preceded by any quantified noun phrase, and sentence (i) contains no sentence which contains 'a', all its anaphors, and all the bound anaphors of any quantified noun phrase appearing in it. Suppose further that sentence (j) is 'a is an A'. Then in any following line (k) one can write the sentence identical to (i) apart from the fact that in it 'some A' has been substituted for 'a'. Line (k) relies on the lines on which lines (i) and (j) rely. Its justification is written 'PI, i, j'.

An example:

1	(1) Wisdom is rare	Premise
2	(2) Wisdom is a virtue	Premise
1,2	(3) Some virtues are rare	PI, 1,2

The constraint, that sentence (1) contain no sentence which contains 'a' and all its anaphors, is needed to block inferences like the one from 'It's not the case that John is tall' and 'John is a man' to 'It's not the case that some men are tall'. It still leaves possible the following inference, though:

1	(1) It's not the case that John is tall	Premise
2	(2) John is a man	Premise
1	(3) John isn't tall	CN/SN, 1
1,2	(4) Some men aren't tall	PI, 3,2

By contrast to 'every A' in Universal Introduction, 'some A' in Particular Introduction need not be substituted for the *only* appearance of 'a' in sentence (i). The fact that this constraint is not necessary is made clear in the validity-preservation proof below. In this way proofs like the following one are possible:

1	(1) Every man likes himself	Premise
2	(2) John is a man	Premise
1,2	(3) John likes himself	UE, 1,2
1,2	(4) John likes John	Anaphora Elimination, 3
1,2	(5) John likes some men	PI, 4,2

Alternatively, we could prove in line (5) the sentence 'Some men like John'.

Validity Preservation. Suppose the sentences on which sentence (k) relies are true. These include the sentences on which sentences (i) and (j) rely, and therefore both are true. Now 'some A' is the first quantified noun phrase in sentence (k), and sentence (k) does not contain any sentence which contains 'some A' and all the bound anaphors of any quantified noun phrase appearing in it. Thus, according to the substitution rule given above (§ 0, p. 109), sentence (k) is true if there is a definite singular noun phrase 'c', such that 'c is an A' is true, and if we substitute 'some A' in sentence (k) by 'c' we get a true sentence. But since sentences (i) and (j) are true, 'a' is such a definite singular noun phrase. So sentence (k) is true, and PI preserves validity.

9.7 Particular Elimination

Suppose that from the fact that a, which is an A, has a certain property, we can prove, without relying on any other fact about a, some general fact, not true specifically of a. Then, since we relied only on the fact that a is an A, what was really significant for our proof is that some A, no matter which, has that property. And what we have proved follows also from the fact that some A has that property.

> *Particular Elimination.* Suppose sentence (i), which does not rely on sentences (j) or (k) and does not contain 'a', contains the noun phrase 'some A', which is not preceded by any other quantified noun phrase, and sentence (i) does not contain any sentence that contains 'some A' and all the bound anaphors of any quantified noun phrase appearing in it. Suppose further that sentence (j) is the premise 'a is an A', and sentence (k) the premise which is identical to sentence (i) apart for the fact that 'a' has been substituted for 'some A'. Now suppose that sentence (l) does not contain a, and does not rely on any sentence which contains a apart from (j) and (k). Then in line (m) sentence (l) can be rewritten, relying on whatever sentences sentences (i) and (l) rely, apart from (j) and (k). It's justification is written 'PE, i, j, k, l'.

An example:

1	(1) Some Athenians are philosophers	Premise
2	(2) All philosophers are wise	Premise
3	(3) Plato is an Athenian	Premise
4	(4) Plato is a philosopher	Premise

2,4	(5) Plato is wise	UE, 2,4
2,3,4	(6) Some Athenians are wise	PI, 5,3
1,2	(7) Some Athenians are wise	PE, 1,3,4,6

Validity Preservation. Suppose the sentences on which sentence (m) relies are true. In that case, all sentences on which sentence (i) relies are true, and sentence (i) is true too. If sentences (j) and (k) were true, then all the sentences on which sentence (l) relies were true, and it were true too. Since sentence (i) is true, then according to the substitution rule given above (§ 0, p. 109), there is a '*c*' so that '*c* is an *A*' is true, and if we substitute 'some *A*' by '*c*' in (i), we get a true sentence. Let us substitute '*a*' by '*c*' in our argument. Since all derivation rules rely only on sameness of definite singular noun phrases and not on the specific definite noun phrase used, the argument up to line (m-1) remains valid. Now since '*c* is an *A*' is true, premise (j) is now true. Moreover, since '*a*' did not appear in (i), (i) remained unchanged after the substitution, and it is still true. But (k) is now the result of substituting '*c*' for 'some *A*' in (i), and is therefore true. And since (j) and (k) are the only premises containing '*a*' on which sentence (l) relies, all other premises on which (l) relies remain true after the substitution. So sentence (l) relies only on true premises, and so sentence (l), that is, sentence (m), is true, and PE preserves validity.

9.8   Referential Import

The two derivation rules for the universal quantifier are compatible with a hypothetical interpretation of it, i.e., with 'Every *A* is *P*' meaning that *if* anything is an *A*, then every *A* is *P*. There might not be any *A* to refer to. According to these derivation rules, 'every *A*' does not have *referential import*. Let us see that in more detail. UE enables us to infer '*a* is a *P*' from (1) 'Every *A* is a *P*' and (2) '*a* is an *A*'; if we read (1) hypothetically, then since *a* is an *A*, some things are *A*'s, and all of them, in particular *a*, are *P*'s. On the other hand, in UI, since the categorical reading, which presupposes reference, entails the hypothetical one, 'every *A*' can also be read hypothetically. We thus need an additional derivation rule to establish the categorical reading.

By contrast, the derivation rules for the particular quantifier, specifically PE, do require a categorical reading of 'Some *A*'s are *P*'; i.e., 'some *A*' presupposes reference to *A*'s. In PE, sentence (l) relies on the premises '*a* is an *A*' and on '*a* is such-and-such', it does not rely on any other premise which contains *a*, and it does not contain '*a*' itself. This makes it possible to substitute '*a*' in the argument, without affecting the argument's validity, by any other expression which preserves the truth of the mentioned premises. That is, sentence (l) actually relies on the fact that some *A*'s are such-and-such. Therefore, to make sentence (l) rely on 'Some *A*'s are such-and-such', without affecting the validity of any argument of this form, the latter sentence should entail the truth of the two premises *for some A*, i.e., it should presuppose that something is an *A*. This referential import of 'some *A*' is expressed in our proof of PE's validity preservation, in that we had to assume that there is some '*c*' such that '*c* is an *A*' is true, and that we get a true sentence if we substitute '*c*' for 'some *A*' in sentence (i). If 'some *A*' could be read hypothetically, that assumption would be false, and PE would not preserve validity.

Let us consider an example, which demonstrates that PE indeed does not preserve validity if 'some *A*' is read hypothetically. The following proof is according to our derivation rules:

1	(1) Some dragons are reptiles	Premise
2	(2) Every reptile is an animal	Premise
3	(3) Sigmund is a dragon	Premise
4	(4) Sigmund is a reptile	Premise
2,4	(5) Sigmund is an animal	UE, 2,4
2,3,4	(6) Some animals are dragons	PI, 3,5
1,2	(7) Some animals are dragons	PE, 1,3,4,6

So according to our derivation rules, 'Some animals are dragons' follows from 'Some dragons are reptiles' and 'Every reptile is an animal'. This indeed is unproblematic, if 'some' and 'every' presuppose reference; in such a case, the truth of the premises entails that of the conclusion. But if we read both hypothetically, we would have a problem. Assume nothing is a dragon. The first premise now means that if anything is a dragon, then some dragons are reptiles. This conditional has a false antecedent, so it is true, or at least it is not false. The second premise now means that if anything is a reptile, then every reptile is an animal; so it is true. The conclusion now means that if anything is an animal, then some animals are dragons; so it is false according to our assumption. We have thus derived a false conclusion from true premises, or at least from premises none of which is false. We thus see that the validity preservation of PE is incompatible with a hypothetical reading of 'some *A*'.

Accordingly, 'Some *A*'s are *P*' has to be read categorically. It follows that if we make 'every' entail 'some', then the hypothetical reading of the former will be ruled out, and it will have to be read categorically. We thus need an additional derivation rule, Referential Import.

> *Referential Import.* Suppose sentence (i) contains the noun phrase 'every *A*', which is not preceded by any other quantified noun phrase, and sentence (i) does not contain any sentence that contains 'every *A*' and all the bound anaphors of any quantified noun phrase appearing in it. Then in any following line (j) one can write the sentence identical to (i) apart from the fact that in it 'some *A*' has been substituted for 'every *A*'. Line (j) relies on the same lines as line (i). Its justification is written 'RI, i'.

An example:

1	(1) John loves every woman	Premise
1	(2) John loves some women	RI, 1

Another example:

1   (1) Peter is a Man   Premise
    (2) Every man is a man   UI, 1,1
    (3) Some men are men   RI, 2

(3) is another theorem of the system developed here, as is any sentence of the form 'Some *A*'s are *A*'s'. Notice that for (3), as well as for (2), the familiar questions arise, should they be considered true, or perhaps just not false, in case there are no *A*'s to refer to? or perhaps there is no general answer, but their truth-value changes with context? or perhaps it is even misguided to try to determine a truth-value in such cases? These questions parallel the same questions for theorems of the predicate calculus of the form, e.g., '(Every *x*)(*Px* or not *Px*)', when applied to an empty domain. These questions are thus not particular to my system, and I need not address them here.

Referential Import does not have any parallel in the derivation rules of the predicate calculus. This is because of the difference between the Universal Elimination rules of the two systems. In the predicate calculus we can infer '*Pa*' from '(Every *x*)*Px*'. By contrast, in natural language we cannot infer '*a* is *P*' from 'Every *A* is *P*'; we need to rely in addition on the sentence '*a* is an *A*'. The referential import of Universal Elimination in natural language relies on the assumption '*a* is an *A*', and not on 'Every *A* is *P*'. In consequence we need an additional derivation rule, to establish the referential import of a noun phrase with a universal quantifier.

As may have been noticed, I talk of a *referential* import of the use of 'every *A*', and not of an *existential* import of such a use, as is often done. That is because the uses of common nouns that we consider presuppose only reference, not existence. For instance, two people can agree that most of the Greek heroes mentioned in the *Iliad* did not return home, while disagreeing as to whether these heroes are real or fictional characters. Their use of 'most of the Greek heroes mentioned in the *Iliad*' presupposes reference, not existence. Accordingly, our derivation rules have to establish that when we talk of every *A* we *refer* to *A*'s, and not that *A*'s exist. (See more on reference to fictional characters and on the presupposition of reference versus that of existence on page 39 above and note 8 there.)

Validity Preservation. According to the substitution rule given above (§ 6.10, p. 109), sentence (i) is true if and only if it is true for *every* substitution of a referring expression which refers to a particular *A* for 'every *A*'. According to the same rule, sentence (j) is true if and only if it is true for at least one such substitution. It follows that the second sentence is true if the first is. But according to RI the two rely on the same sentences. So if the premises of sentence (j) are true, so are those of sentence (i), so sentence (i) is true, and so sentence (j) is also true. So RI preserves validity.

Referential Import was the last derivation rule of my system, and it was proved, as were all other rules, to preserve validity. In addition, a one-line argument is always valid, since its only line is a premise. We have thus proved by induction that any argument is valid, and my system is consistent.

Chapter 10

Applications I: Aristotelian Logic

Having given the rules of my system, I now proceed to apply it to natural language. I first show that all the formal logical relations recognized by Aristotelian logic and expressible in my system are derivable by means of these rules.[1]

10.1   The Square of Opposition

Aristotelian logic occupied itself almost exclusively with the logical relations between sentences of the form 'Every *A* is/isn't a *B*' and 'Some *A*'s are/aren't *B*'s'. The inferences constructed out of these sentences were divided into three groups: the Square of Opposition, immediate inferences, and syllogisms. In this and the following two sections I shall show how all these inferences are derivable within my system.

I start with the Square of Opposition. It consists of inferences with a single premise and a conclusion that has the same subject and predicate as the premise. The Square of Opposition includes the following six logical relations:

1. 'Every *A* is a *B*' entails 'Some *A*'s are *B*'s'.
2. 'Every *A* isn't a *B*' entails 'Some *A*'s aren't *B*'s'.
3. 'Every *A* is a *B*' and 'Some *A*'s aren't *B*'s' are contradictory.
4. 'Every *A* isn't a *B*' and 'Some *A*'s are *B*'s' are contradictory.
5. 'Every *A* is a *B*' is contrary to 'Every *A* isn't a *B*' (it cannot be that both are true).
6. 'Some *A*'s are *B*'s' is sub-contrary to 'Some *A*'s aren't *B*'s' (it cannot be that both are false).

(1) and (2) are proved by direct application of Referential Import. Let us prove (3):

1	(1) Every *A* is a *B*	Premise
2	(2) Some *A*'s aren't *B*'s	Premise
3	(3) *a* is an *A*	Premise
4	(4) *a* isn't a *B*	Premise
4	(5) It's not the case that *a* is a *B*	CN/SN, 4
1,3	(6) *a* is a *B*	UE, 1,3

---

[1] In my presentation of Aristotelian logic I rely mainly on Bergmann (1975) and Strawson (1952).

3,4 (7) It's not the case that every *A* is a *B*   NI (PCDR), 1,5,6
2   (8) It's not the case that every *A* is a *B*   PE, 2,3,4,7

('NI' abbreviates 'Negation Introduction'.) Since 'Some *A*'s aren't *B*'s' entails the negation of 'Every *A* is a *B*', it follows by the rules of the propositional calculus that 'Every *A* is a *B*' also entails the negation of 'Some *A*'s aren't *B*'s'. We should now prove that the negation of 'Every *A* is a *B*' entails 'Some *A*'s aren't *B*'s'. To save space, I shall write 'Not (*p*)' for 'It's not the case that *p*'.

1   (1) Not (every *A* is a *B*)   Premise
2   (2) Not (some *A*'s aren't *B*'s)   Premise
3   (3) *a* is an *A*   Premise
4   (4) *a* isn't a *B*   Premise
3,4   (5) Some *A*'s aren't *B*'s   PI, 4,3
2,3   (6) Not (*a* isn't a *B*)   NI (PCDR), 5,2,4
2,3   (7) Not (Not (*a* is a *B*))   CN/SN, 6
2,3   (8) *a* is a *B*   NE (PCDR), 7
2   (9) Every *A* is a *B*   UI, 8,3
1   (10) Not (Not (some *A*'s aren't *B*'s)) NI (PCDR), 9,1,2
1   (11) Some *A*'s aren't *B*'s   NE (PCDR), 10

('NE' abbreviates 'Negation elimination'.) Again, since the negation of 'Every *A* is a *B*' entails 'Some *A*'s aren't *B*'s', it follows by the rules of the propositional calculus that the negation of 'Some *A*'s aren't *B*'s' entails 'Every *A* is a *B*'. So (3) is proved. (4) is established by similar arguments.

Since 'Every *A* is a *B*' entails 'Some *A*'s are *B*'s' (by 1), which entails the negation of 'Every *A* isn't a *B*' (by 4), and since the same is true, with suitable modifications, in the opposite direction, (5) is also established.

Lastly, let us prove (6) above, that 'Some *A*'s are *B*'s' and 'Some *A*'s aren't *B*'s' are sub-contraries. That is proved since the negation of 'Some *A*'s are *B*'s' entails 'Every *A* isn't a *B*' (by 4), and the latter entails 'Some *A*'s aren't *B*'s' (by 2), and similarly in the other direction.

We have thus proved all the logical relations of the Aristotelian Square of Opposition. By contrast, on every acceptable translation of the Square's sentences into the predicate calculus, some of its relations turn out invalid. Now the Square of Opposition is intuitively valid, as is testified by its universal acceptance by logicians from the time of Aristotle until the twentieth century. Thus, this greater success of my system demonstrates its superiority over the predicate calculus in capturing the logic of natural language. The same is demonstrated by the greater success of my system compared with the predicate calculus in establishing the validity of immediate inferences and syllogisms, which will be proved in the next two sections.

10.2 Immediate Inferences

Aristotelian logic classified as immediate inferences those inferences that have a single premise and in which the subject or the predicate of the conclusion are different from those of the premise. Four kinds of immediate inferences were recognized: conversion, obversion, contraposition and inversion.

The last three kinds – obversion, contraposition and inversion – involve as subject or predicate of the conclusion the negation of either the subject or the predicate of the premise. For instance, one was supposedly able to infer from the premise 'Every man is mortal' the conclusion 'Every non-mortal is non-man'. But the very coherence of negated concepts such as 'non-man' is dubious, especially when they are supposed to be used referentially. Moreover, I did not discuss the logic of negated terms in this book. Inferences of these kinds will therefore not be considered here.

Conversion, by contrast, contained the following three inferences (cf. Aristotle, *Prior Analytics*, Book I, Chapter 2):

1. 'Every A isn't a B' entails 'Every B isn't an A'.
2. 'Every A is a B' entails 'Some B's are A's'.
3. 'Some A's are B's' entails 'Some B's are A's'.

The last two inferences can be deduced from the first together with the laws of the Square of Opposition (Bergmann, 1975, 282-3). I shall therefore prove only the first conversion.

1	(1) Every A isn't a B	Premise
2	(2) a is a B	Premise
3	(3) a is an A	Premise
1,3	(4) a isn't a B	UE, 1,3
1,3	(5) Not(a is a B)	CN/SN, 4
1,2	(6) Not (a is an A)	NI (PCDR), 2,5,3
1,2	(7) a isn't an A	CN/SN, 6
1	(8) Every B isn't an A	UI, 7,2

All conversions are therefore provable by means of the system developed above.

Conversions presuppose that the concept which appears as a predicate in the premise – 'B' in our formulations above – can be used as a referring expression as well: that is how it is used in the conclusions. This is not always the case, as I have tried to show above (§ 2.3). For instance, from 'Some people are tall' one cannot infer neither 'Some tall are people' nor 'Some tall things are people'. The first sentence is not even grammatical; and the concept 'tall' is attributive, and so it cannot be used without an appropriate additional referring concept to refer to particulars, as it is supposed to be used in the second sentence. Now, the derivation rules presented in the previous chapter do not distinguish between concepts that can be used both as predicates and as referring expressions and concepts that can be used only as predicates. Formal logic presupposes that the sentences it manipulates make sense semantically. The avoidance of conclusions which are meaningless – not simply false! – is partly left to non-formal considerations.

10.3 Syllogisms

Syllogisms are inferences with two premises, in which the subject of the conclusion is either the subject or the predicate of one of the premises, the predicate of the conclusion is either the subject or predicate of the other premise, and both premises contain one more concept, the middle term, as either the subject or predicate of each. Syllogisms were traditionally divided into four groups, four *figures*, each containing several valid inferences. It was shown that all valid syllogisms of the second, third and fourth figures are derivable, by means of the rules of the Square of Opposition and immediate inferences, from the valid syllogisms of the first figure (direct reductions), and that all these are further reducible by the same means together with equivalencies of the propositional calculus to the first syllogism of the first figure, namely to:

> Every A is a B
> Every B is a C
> ∴Every A is a C [2]

Accordingly, to show that all valid syllogisms are derivable within my system, it would suffice to show that this one, traditionally called *Barbara*, is so derivable. This is done as follows:

1	(1) Every A is a B	Premise
2	(2) Every B is a C	Premise
3	(3) a is an A	Premise
1,3	(4) a is a B	UE, 1,3
1,2,3	(5) a is a C	UE, 2,4
1,2	(6) Every A is a C	UI, 5,3

Although for the reasons just mentioned the proof of Barbara is sufficient for my aims, I shall also prove the other three valid syllogisms of the first figure mentioned by Aristotle (*Prior Analytics*, Book I, Chapter 4), namely, Celarent, Darii and Ferio. Firstly, Celarent:

1	(1) Every A is a B	Premise
2	(2) Every B isn't a C	Premise
3	(3) a is an A	Premise
1,3	(4) a is a B	UE, 1,3
1,2,3	(5) a isn't a C	UE, 2,4
1,2	(6) Every A isn't a C	UI, 5,3

---

[2] See Strawson, 1952, pp. 158-63.

Secondly, Darii:

1	(1) Some *A*'s are *B*'s	Premise
2	(2) Every *B* is a *C*	Premise
3	(3) *a* is an *A*	Premise
4	(4) *a* is a *B*	Premise
2,4	(5) *a* is a *C*	UE, 2,4
2,3,4	(6) Some *A*'s are *C*'s	PI, 5,3
1,2	(7) Some *A*'s are *C*'s	PE, 1,3,4,6

Lastly, Ferio:

1	(1) Some *A*'s are *B*'s	Premise
2	(2) Every *B* isn't a *C*	Premise
3	(3) *a* is an *A*	Premise
4	(4) *a* is a *B*	Premise
2,4	(5) *a* isn't a *C*	UE, 2,4
2,3,4	(6) Some *A*'s aren't *C*'s	PI, 5,3
1,2	(7) Some *A*'s aren't *C*'s	PE, 1,3,4,6

This concludes what I set out to prove in this chapter, i.e., that all formal logical relations recognized by Aristotelian logic and expressible by my system are provable in it. As I said above, I think this success supports my claim that my logical system, and the semantic analysis on which it is grounded, express the semantic and logic of natural language more accurately than does the predicate calculus.

Chapter 11

Applications II:
Beyond Aristotelian Logic

I now proceed to apply my system to logical relations which were not considered by Aristotelian logic. These will include logical relations between multiply quantified sentences; logical relations between propositional combinations of quantified sentences; and some logical relations between relations. I shall also consider how identity should be incorporated in my system. I precede these topics by a generalization of Transposition.

11.1 Generalization of Transposition

Transposition was formulated above for predicates with singular definite noun phrases as subjects. It can be generalized, however, to any predicate one of whose subjects is a quantified noun phrase of the form 'every A' or 'some A', the other subjects being definite singular noun phrases.

Instead of proving the generalized form of Transposition for any predicate, which would require using complex symbols and indices, I shall prove it for two examples, one with 'every' as quantifier, the other with 'some'. The generalization of this proof to any predicate is straightforward.

Firstly, the universal quantifier:

1	(1) John gave the syllabus to every student	Premise
2	(2) Peter is student	Premise
1,2	(3) John gave the syllabus to Peter	UE, 1,2
1,2	(4) The syllabus was given to Peter by John	T, 3
1	(5) The syllabus was given to every student by John	UI, 4,2

Secondly, the particular quantifier, 'some':

1	(1) John gave the syllabus to some students	Premise
2	(2) Peter is student	Premise
3	(3) John gave the syllabus to Peter	Premise
3	(4) The syllabus was given to Peter by John	T, 3
2,3	(5) The syllabus was given to some students by John	PI, 4,2
1	(6) The syllabus was given to some students by John	PE, 1,2,3,5

Although I could have formulated Transposition to begin with in its generalized form, explaining why it preserves validity, I chose not to do that. There is an obvious logical advantage in minimizing one's principles, showing how all other necessary rules are derivable from them.

One of the cases to which the generalized form of Transposition proved above can be applied is that in which the only quantified noun phrase in a sentence is preceded by more than a single definite singular noun phrase. The possible transpositions in that case include those in which the definite singular noun phrases preceding the quantified noun phrase are transposed in any possible manner, while the order of the quantified noun phrase and all noun phrases following it remains unchanged. I shall now use this result to prove that for any predicate, if its quantified noun phrases are preceded by more than a single definite singular noun phrase, these definite singular noun phrases can be transposed in any possible manner, leaving the order of all other noun phrases unchanged. This result will be used below. I shall prove it by induction on the number of quantified noun phrases in the sentence.

We have already proved the result for the case of a single quantified noun phrase. Suppose now that the claim is true for any predicate including n quantified noun phrases among its subjects, and let us prove it for $n+1$.

Suppose sentence S is 'np_1, ... np_m is/isn't P', and that it contains $n+1$ quantified noun phrases, and let us prove the claim for it. Suppose the first quantified noun phrase in S is of the form 'every A'. We can then write 'a is an A' as a premise (where 'a' does not appear in S), and then substitute in the next line 'a' for 'every A' by means of UE. We can then, according to our inductive supposition, transpose the definite singular noun phrases preceding 'a' in any way we wish. Finally, we can substitute 'every A' for 'a', relying on UI. This proves the claim for the case considered here. A proof similar to the one for 'some' in our generalization of Transposition will apply in case the first quantified noun phrase in S is of the form 'some A'. This completes the inductive proof.

11.2 Multiply Quantified Sentences

In this section I prove some relations between multiply quantified sentences. I chose some central cases in order to demonstrate the power of my system and how it is applied.

I first show that the order of two universally quantified noun phrases, as well as the order of two particularly quantified noun phrases, which are the subjects of a two-place predicate, can be changed. Firstly, the universal quantifier:

1	(1) Every man loves every woman	Premise
2	(2) John is a man	Premise
3	(3) Mary is a woman	Premise
1,2	(4) John loves every woman	UE, 1,2
1,2,3	(5) John loves Mary	UE, 4,3

1,2,3	(6) Mary is loved by John	T, 5
1,3	(7) Mary is loved by every man	UI, 6,2
1	(8) Every woman is loved by every man	UI, 7,3

Secondly, the particular quantifier:

1	(1) Some men love some women	Premise
2	(2) John is a man	Premise
3	(3) John loves some women	Premise
4	(4) Mary is a woman	Premise
5	(5) John loves Mary	Premise
5	(6) Mary is loved by John	T, 5
2,5	(7) Mary is loved by some men	PI, 6,2
2,4,5	(8) Some women are loved by some men	PI, 7,4
2,3	(9) Some women are loved by some men	PE, 3,4,5,8
1	(10) Some women are loved by some men	PE, 1,2,3,9

These two proofs are *in*sufficient to prove the counterpart in this system of the laws of the predicate calculus, that the sentences '$\forall x \forall y F(x,y)$' and '$\forall y \forall x F(x,y)$' are equivalent, as well as the sentences '$\exists x \exists y F(x,y)$' '$\exists x \exists y F(x,y)$'. This is because in the predicate calculus, the two quantifiers can be followed by other quantifiers, while for my use of Transposition in the proofs above it was necessary that the sentence does not contain any additional quantified noun phrase. However, I proved in the previous section that Transposition can be generalized to cases where any number of quantified noun phrases are preceded by definite singular noun phrases which are to be transposed. In this way the possibility to transpose, in any predication, two universally or two particularly quantified noun phrases that are not preceded by other quantified noun phrases can be generally proved. One should simply use the generalized form of Transposition in the sixth line of each of the above proofs.

I now proceed to prove that if a two-place predicate has 'some' as the quantifier of its first subject and 'every' as the quantifier of its second subject, then their order can be transposed.

1	(1) Some women are loved by every man	Premise
2	(2) Jane is a woman	Premise
3	(3) Jane is loved by every man	Premise
4	(4) Peter is a man	Premise
3,4	(5) Jane is loved by Peter	UE, 3,4
3,4	(6) Peter loves Jane	T, 5
2,3,4	(7) Peter loves some women	PI, 6,2
2,3	(8) Every man loves some women	UI, 7,4
1	(9) Every man loves some women	PE, 1,2,3,8

Considerations paralleling those given above for the case of two consecutive universal or particular quantifiers enable the generalization of this proof to any predicate with any number of quantified noun phrases as its subjects: If 'some *A*' is followed by 'every *B*' as subjects of a predicate, without any quantified noun phrase coming between them or preceding them, they can be transposed.

This proof does not work the other way around, i.e., in case the first subject has 'every' as its quantifier while the second one has 'some' as its. The reason is that in particular elimination, the quantified noun phrase should not be preceded by any other quantified noun phrase. Let us see that in detail:

1	(1)Every man loves some women	Premise
2	(2) Jane is a woman	Premise
3	(3) Every man loves Jane	Premise
4	(4) Peter is a man	Premise
3,4	(5) Peter loves Jane	UE, 3,4
3,4	(6) Jane is loved by Peter	T, 5
3	(7) Jane is loved by every man	UI, 6,4
2,3	(8) Some women are loved by every man	PI, 7,2

But now we cannot make (8) rely on (1) instead of (2) and (3), since 'some women' is not the first quantified noun phrase in (1). (8) is true because (3) is true for a particular woman, but (1) need not be true for any particular woman. This demonstrates the necessity for the consistency of my system of the condition in PE that 'some *A*' will not be preceded by any other quantified noun phrase.

My proofs in this section concerned various cases of transposition of two quantified noun phrases that are subjects of the same predicate. This is not the exact parallel of changing quantifier order in the predicate calculus: in the predicate calculus not only a single predicate, but any sentential function, can be in the scope of the quantifiers. But this general possibility has no exact parallel in natural language. In natural language, if the two quantified noun phrases are not subjects of the same predicate, then changing their order in the sentence will involve changing the order in the sentence of the predicates of which they are subjects. And the possibility of the latter change of order depends on equivalencies of the propositional calculus (e.g., '*p* & *q*' and '*q* & *p*') and on the capability of natural language to express the same propositional formulas in different ways (e.g., 'If *p*, *q*' and '*q* if *p*'). There is thus in this case no general problem that belongs to the theory of quantification. I shall, however, demonstrate in the next section the possibility of some changes of order of quantified noun phrases in propositional combinations of predications.

11.3   Predicate- and Sentence-Connectives

In section 8.5 we saw that natural language, perhaps in order to avoid ambiguities, prefers in many cases the use of connected predicates over the use of bound anaphora. For instance, we say 'Some men are *tall and handsome*', if we want to attribute two properties to a number of men. We cannot, apparently, say the same thing by the use of bound anaphora: in the sentence 'Some men are tall, and they are also handsome' the pronoun 'they' is not a bound anaphor but an E-type pronoun. The second conjunct presupposes the truth of the first, and 'they' is used to denote those men who are tall.

For that reason, a deductive system that attempts to analyze the entailment relations between sentences of natural language, has to consider the relations between predicate-connectives and sentence-connectives. We should be able, for instance, to derive 'John is tall and John is fat' from 'John is tall and fat', and *vice versa*. I shall make a few steps in that direction in this section. The rules I shall formulate below resemble the substitution rule for sentence- and copula-negation, itself a rule relating a sentence-connective to a manner of predication.

First, predicate conjunction. I call this rule 'PC/SC', for *Predicate Conjunction – Sentence Conjunction*.

> *PC/SC*. Suppose sentence (i) is or contains the sentence '*a* is *A* and *a* is *B*'. Then in any following line (j) the sentence identical to sentence (i), but with '*a* is *A* and *B*' substituted for '*a* is *A* and *a* is *B*', can be written. And if sentence (i) is or contains the sentence '*a* is *A* and *B*', then in any following line (j) the sentence identical to sentence (i), but with '*a* is *A* and *a* is *B*' substituted for '*a* is *A* and *B*', can be written. In both cases sentence (j) relies on the same premises as sentence (i), and its justification is written 'PC/SC, i'.

Secondly, predicate disjunction. I shall call this rule 'PD/SD', for *Predicate Disjunction – Sentence Disjunction*.

> *PD/SD*. Suppose sentence (i) is or contains the sentence '*a* is *A* or *a* is *B*'. Then in any following line (j) the sentence identical to sentence (i), but with '*a* is *A* or *B*' substituted for '*a* is *A* or *a* is *B*' can be written. And if sentence (i) is or contains the sentence '*a* is *A* or *B*', then in any following line (j) the sentence identical to sentence (i), but with '*a* is *A* or *a* is *B*' substituted for '*a* is *A* or *B*', can be written. In both cases sentence (j) relies on the same premises as sentence (i), and its justification is written 'PD/SD, i'.

An example with predicates disjunction:

1	(1) Every *A* is either *B* or *C*	Premise
2	(2) Some *A*'s aren't *B*'s	Premise
3	(3) John isn't *B*	Premise
4	(4) John is an *A*	Premise
1,4	(5) John is either *B* or *C*	UE, 1,4
1,4	(6) John is *B* or John is *C*	PD/SD, 5
3	(7) It's not the case that John is *B*	CN/SN, 3
1,3,4	(8) John is *C*	DS (PCDR), 6,7
1,3,4	(9) Some *A*'s are *C*'s	PI, 8,4
1,2	(10) Some *A*'s are *C*'s	PE, 2,4,3,9

An example with predicates conjunction:

1	(1) Some men are Greek and mortal	Premise
2	(2).Socrates is a man	Premise
3	(3) Socrates is Greek and mortal	Premise
3	(4) Socrates is Greek and Socrates is mortal	PC/SC, 3
3	(5) Socrates is mortal	&E (PCDR), 4
2,3	(6) Some men are mortal	PI, 5,2
1	(7) Some men are mortal	PE, 1,2,3,6

Another example:

1	(1) Not(Some men are A and B)	Premise
2	(2) Some men are A	Premise
3	(3) Socrates is a man	Premise
4	(4) Socrates is A	Premise
5	(5) Socrates is B	Premise
4,5	(6) Socrates is A and Socrates is B	&I (PCDR), 4,5
4,5	(7) Socrates is A and B	PC/SC, 6
3,4,5	(8) Some men are A and B	PI, 3,7
1,3,4	(9) Not(Socrates is B)	NI (PCDR), 5,1,8
1,3,4	(10) Socrates isn't B	CN/SN, 9
1,3,4	(11) Some men aren't B	PI, 10,3
1,2	(12) Some men aren't B	PE, 2,3,4,11

One last example:

1	(1).John is tall and he is also handsome	Premise
2	(2).John is a man	Premise
1	(3).John is tall and handsome	PC/SC, 1
1,2	(4).Some men are tall and handsome	PI, 3,2

We needed to change the sentence conjunction in (1) into predicate conjunction in (3) in order to introduce the particular quantifier, since it cannot bind an anaphor across sentence conjunction.

Of course, in some cases bound anaphora across sentence-connectives can be used. For such cases, the derivation rules brought in the preceding sections should suffice. One example with anaphoric noun phrases and implication was brought above, on page 120. I shall give here two other examples, demonstrating the possibility of changing the order of quantified noun phrases in such sentences.

1	(1)If a man is a vet and a donkey passes by, he vaccinates it	Premise
2	(2)John is a man	Premise
3	(3)Platero is a donkey	Premise
1,2	(4)If John is a vet and a donkey passes by, he vaccinates it	UE, 1,2
1,2,3	(5)If John is a vet and Platero passes by, he vaccinates it	UE, 4,3
1,2,3	(6)Platero is vaccinated by John, if he is a vet and it passes by	T, 5
1,3	(7)Platero is vaccinated by a man, if he is a vet and it passes by	UI, 6,2
1	(8)A donkey is vaccinated by a man, if he is a vet and it passes by	UI, 7,3

Second example:

1	(1) Some philosophers are admired by any man, if he appreciates philosophy	
		Premise
2	(2) Plato is admired by any man, if he appreciates philosophy	Premise
3	(3) Plato is a philosopher	Premise
4	(4) John is a man	Premise
2,4	(5) Plato is admired by John, if he appreciates philosophy	UE, 2,4
2,4	(6) If John appreciates philosophy, then he admires Plato	T, 5
2,3,4	(7) If John appreciates philosophy, then he admires some philosophers	
		PI, 6,3
2,3	(8) If a man appreciates philosophy, then he admires some philosophers	
		UI, 7,4
1	(9) If a man appreciates philosophy, then he admires some philosophers	
	PE, 1,2,3,8	

In both arguments I tacitly relied – in line (6) in both cases – on the synonymy in natural language, at least where no quantification is involved, of sentences of the form 'If *p* then *q*' and '*q*, if *p*'.

The next argument I shall consider in this section is the following:

> Every philosopher is a human-being; hence, every hat of a philosopher is a hat of a human-being.

This argument is interesting not so much because it is about philosophers' hats, but because already De Morgan has shown (1847, p. 114) that it is not reducible to the Aristotelian syllogism. We should see whether our system fares any better. However, since we did not discuss the logic of noun clauses such as 'hat of a philosopher' – nouns with defining clauses that contain quantified noun phrases – I shall have to paraphrase the conclusion in order to prove it. (The predicate calculus also paraphrases such predicates in order to translate them.) The paraphrase I shall prove is 'If a hat belongs to a philosopher, it belongs to a human being'. This conditional sentence is not perfectly synonymous with the categorical it paraphrases, though, since only the latter presupposes reference to philosophers' hats.

1	(1) Every philosopher is a human-being.	Premise
2	(2) This is a hat	Premise
3	(3) This belongs to a philosopher	Premise
4	(4) This belongs to Plato	Premise
5	(5) Plato is a philosopher	Premise
1,5	(6) Plato is a human-being	UE, 1,5
1,4,5	(7) This belongs to a human-being	PI, 4,6
1,3	(8) This belongs to a human being	PE, 3,4,5,7
1	(9) If this belongs to a philosopher, it belongs to a human being	
		→I (PCDR), 3,8

1   (10) If a hat belongs to a philosopher, it belongs to a human being
                                                                        UI, 9,2

The method of paraphrase utilized while dealing with this argument is
generally applicable when a premise or conclusion contains a noun phrase with a
defining relative clause, itself containing a quantified construction. Consider, for
instance, the following argument:

> Every *A* who is a *B* is a *C*
> Some *A*'s are *B*'s or *C*'s
> ∴ Some *A*'s are *C*'s

To prove this argument we should paraphrase the first premise by 'If an *A* is a *B*
then it is a *C*':

1	(1) If an *A* is a *B* then it is a *C*	Premise
2	(2) Some *A*'s are *B*'s or *C*'s	Premise
3	(3) Plato is an *A*	Premise
4	(4) Plato is a *B* or a *C*	Premise
1,3	(5) If Plato is a *B* then he is a *C*	UE, 1,3
4	(6) Plato is a *B* or Plato is a *C*	PD/SD, 4
1,3,4	(7) Plato is a *C*	PCDR, 5,6
1,3,4	(8) Some *A*'s are *C*'s	PI, 7,3
1,2	(9) Some *A*'s are *C*'s	PE, 2,3,4,8

11.4 The Logic of Relations

I shall now use my system to prove a few logical relations between relations. These
examples are needed in order to show how my system should be applied to such
cases. Moreover, since I do not prove the completeness of my system, these
examples also demonstrate the system's deductive power.

   First, let us prove that an asymmetric relation is irreflexive. I shall prove it for a
specific asymmetric relation, *older than*.

1	(1) If any man is older than any other man, then the second man isn't older than the first	Premise
2	(2) John is a man	Premise
1,2	(3) If John is older than any other man, then the second man isn't older than John	UE, 1,2
1,2	(4) If John is older than John, then John isn't older than John	UE, 3,2
1,2	(5) If John is older than John, then it's not the case that John is older than John	CN/SN, 4
1,2	(6) It's not the case that John is older than John	PCDR, 5
1,2	(7) John isn't older than John	CN/SN, 6
1	(8) Every man isn't older than himself	UI, 7,2

Next, I shall prove that any intransitive relation is irreflexive. I shall prove it for the intransitive relation *parent of*. I allow myself a few shortcuts.

1	(1) If a person x is a parent of a person y, and person y is a parent of a person z, then person x isn't a parent of person z	Premise
2	(2) Jane is a person	Premise
1,2	(3) If Jane is a parent of Jane, and Jane is a parent of Jane, then Jane isn't a parent of Jane	UE, 1,2
1,2	(4) If Jane is a parent of Jane, and Jane is a parent of Jane, then it's not the case that Jane is a parent of Jane	CN/SN, 3
1,2	(5) It's not the case that Jane is a parent of Jane	PCDR, 4
1,2	(6) Jane isn't a parent of Jane	CN/SN, 5
1	(7) Every person isn't a parent of himself	UI, 6,2

I used in sentence (1) the locutions 'a person x' and 'person x' as a universally quantified noun phrase and its bound anaphor, respectively. These and similar locutions, although having a technical or formal ring, are often used in spoken and written natural language when claims as those made in sentence (1) – themselves having the same ring – are made. I therefore found it legitimate to use them in my arguments, which are supposed to be carried out in natural language, minimally stylized.

Lastly, let us prove that any symmetric and transitive relation is also reflexive, if any particular relates to at least one particular. I shall again prove this for a particular relation, this time the relation *as old as*. And again, I shall allow myself a few shortcuts.

1	(1)	If a man x is as old as a man y, then man y is as old as man x	Premise
2	(2)	If a man x is as old as a man y, and man y is as old as a man z, then man x is as old as man z	Premise
3	(3)	Every man is as old as some men	Premise
4	(4)	John is a man	Premise
3,4	(5)	John is as old as some men	UE, 3,4
6	(6)	Peter is a man	Premise
7	(7)	John is as old as Peter	Premise
2,4,6	(8)	If John is as old as Peter, and Peter is as old as John, then John is as old as John	UE, 2,4,6
1,4,6	(9)	If John is as old as Peter, then Peter is as old as John	UE, 1,4,6
1,4,6,7	(10)	Peter is as old as John	MP (PCDR), 9,7
1,2,4,6,7	(11)	John is as old as John	&I+MP (PCDR), 7,10,8
1,2,3,4	(12)	John is as old as John	PE, 5,6,7,11
1,2,3	(13)	Every man is as old as himself	UI, 12,4

11.5   Identity

The last subject I shall discuss in this chapter is how the semantic principles developed in this work should be applied to the analysis of identity, and how identity should be incorporated in the deductive system developed here.

A typical sentence used to make an identity statement is the following:

John is the man standing over there.

Grammatically, this sentence resembles a subject-predicate sentence, such as 'John is tall' or 'John is a man'. *Prima facie*, identity statements are made with sentences in which the predicate is a singular referring expression. And in fact, I do not see why they should not be considered such sentences. Indeed, the nature of predication in such sentences is different from its nature in other sentences. But then, there is very little in common between other kinds of predication either. Consider, for instance, the following examples:

John is 1.85 meters tall
John is tall
John is strong
John is intelligent
John is asleep
John is a man
John is an engineer.

The nature of predication in any one of these sentences is very different from its nature in any other one. Specifying the dimensions of somebody, comparing them to common dimensions, ascribing an ability, ascribing something which resembles a complex ability, describing one's state, classifying, telling one's occupation: all these have very little in common. The common semantic content of acknowledged forms of predication is minimal; it is merely saying something about the particulars referred to. And when we say that John is *the man standing over there*, we also say something about John: we say who he is.

What *does* distinguish the predication of identity from all the examples just given is that all other predicates can apply to more than a single individual; while if John is the man standing over there, then no one else is the man standing over there. But this uniqueness in application is true of some other acknowledged predicates as well, e.g., 'won the race' or 'the tallest man in Oxford'. In fact, predication as in the last example, where the predicate is a definite description, shades into identity statements. So the number of particulars to which a predicate can apply does not constitute a reason against considering identity statements subject-predicate in structure.

The difference between my analysis of identity and Frege's is a result, and in a way epitomizes, our different conceptions of predication. Frege thought that concept-words are always predicative, designating concepts or functions from arguments to truth-values. According to him, what is characteristic of a concept is its unsaturatedness (*Ungesättigtheit*), its need of supplementation or its predicative nature (1891, p. 17; 1892, pp. 193, 197, 205). This unified analysis of concepts he took to be one of his main achievements, as is attested by a note written in August 1906, entitled 'Was kann ich als Ergebnis meiner Arbeit ansehen?'. This analysis of concepts does not leave any room for construing identity sentences as subject-*predicate* in nature (1892, p. 194).

By contrast to Frege, I maintain that concepts, or what he called 'concept-words', can also be used as referring expressions, a use in which they are as little "unsaturated" as proper names are; while I consider predication semantically heterogeneous, having in common only the minimal content of *saying something about*. On this approach, the analysis of singular referring expressions in identity sentences as predicates is a natural one.

Ironically, Frege was mislead into a mistaken view of concepts, a view which he considered one of his main achievements, by what he time and again warned against: mistaking mere grammatical uniformity for a logical or semantic one. The semantic diversity of predication, acknowledged by logicians from Aristotle's *Categories* on, disappeared under Frege's pseudo-homogenous semantic relation of falling under a concept.

My minimalist conception of predication is close to that of Aristotle. Following Plato in the *Sophist* (262c-263d), Aristotle maintained that a simple proposition is composed of a noun (*onoma*) and a verb (*rhema*). Of the verb he says that 'it is always a sign of something said of something else, i.e. of something either predicable of or present in some other thing' (*On Interpretation* 3, 16b10). This minimalist conception seems to agree with that of the Stoics as well: according to Apollodorus and his followers, a predicate is what is said of something, or a thing associated with one or more subjects (DL vii.64).

There has always been some dissatisfaction with the analysis of identity as a relation, one holding necessarily between any thing and itself alone. In consequence, Frege (1879, § 8) and others have repeatedly attempted to interpret identity statements, especially those involving two proper names, as statements about words – attempts which were never satisfactory. If identity is a relation, it is certainly an exceptional one, one which probably extends the meaning of 'relation'. The analysis of identity as a relation is not any more plausible than its analysis as a kind of predication.

Moreover, if we consider identity statements a kind of subject-predicate statements, then an alleged ambiguity of the copula is dismissed as only apparent and a result of a mistaken semantic analysis. The 'is' of 'John is tall' and that of 'John is the man standing over there' are, semantically, one and the same. The 'is' of identity is the 'is' of predication. This simplifying result supports, of course, our analysis. In addition, if these uses of the copula were semantically different, then the fact that the same ambiguity exists in many remote languages would be an improbable coincidence.[1]

We find the idea that identity is a relation natural for several reasons. This is how we have seen identity construed ever since we started learning logic, we have seen this idea at work within a powerful deductive system, and identity has been assimilated – in notation and function – to the relation of mathematical equality (cf. Frege, 1892, p. 194). What we find natural is typically what we are accustomed to.

But other logicians, familiar with different theories, were differently disposed. For instance, according to Al-Fārābī, the great Muslim logician and philosopher (c.870-950), a singular referring expression – an expression signifying an individual, in his terminology – can be a predicate in a subject-predicate judgment (1956, p. 128, § 1). His examples were 'Zaid is this one standing' and 'This one standing is Zaid' (ibid., § 2).

Similarly, Aquinas writes about 'propositions in which the same thing is predicated of itself' (*Summa Theologiae* I, Question 13, Article 12, c) – meaning, presumably, propositions like 'Socrates is Socrates'. It is obvious to him that the name 'Socrates' functions as a predicate in its second appearance in this sentence. What he tries to explain is how the term changes its function from subject to predicate.[2]

Free of contingent cultural inclinations, there is no obstacle to considering identity statements a kind of subject-predicate statement.

It seems, in fact, that no one previous to Frege has maintained that the copula is ambiguous in this respect, and that identity sentences are not subject-predicate ones. From a historical perspective, the conviction of the philosophers and logicians of the last century that this ambiguity is the case is the exception in need of justification, and not the other way round.

And there have been a few dissenting voices even in our Fregeian era. Fred Sommers (1969) also maintained that the 'is' of identity is that of predication, and for reasons similar to some of mine: 'not proliferating senses of "is" beyond necessity' (p. 500) and avoiding 'Frege's famous difficulty with identity statements' (p. 504).[3]

---

[1] There is therefore, in my opinion, some ground for challenging Russell's claim that 'it is a disgrace to the human race that it has chosen to employ the same word "is" for these entirely different ideas' (1919, p. 172); but this should be the subject of a separate work.

[2] Since his solution (ibid.) is not strictly relevant to my discussion in the text, but may still be of interest, I bring it in a footnote. Aquinas thinks that 'intellect treats what it assigns to the subject as being on the side of the referent, whereas it treats what it assigns to the predicate as belonging to the nature of a form existing in the referent, in accordance with the saying that predicates are taken formally and subjects materially.' The reference and translation are taken, with some modification, from Weidemann (1986, pp. 187-8).

[3] Sommers, however, admits noun phrases of the form 'all Socrates' and 'some Socrates' into his system, in order to account for inferences involving proper names, even in sentences that are not identity-sentences. But surely sentences like 'All Socrates is (or are) unwise', 'Some Socrates is Socrates' etc., where 'Socrates' functions as a proper name, are not part of natural language. Sommers' logic, even if it yields valid arguments, cannot be considered as a logic of

He does not mention, however, anything like my minimalist conception of predication, which is my main reason for considering identity sentences a form of subject-predicate ones.

Independently of Sommers, Michael Lockwood, inspired by Mill, has also maintained that proper names can function as predicates. Mill thought that proper names have denotation but no connotation (Mill, 1872, 1.II.5, p. 31), and that the meaning of a subject-predicate sentence, where the subject is a proper name, 'is, that the individual thing denoted by the subject, has the attributes connoted by the predicate' (ibid., 1.V.4, p. 108). Mill consequently maintained that when we predicate of anything its proper name we convey only that this is its name (ibid., 1.II.5, p. 37; cf. ibid., p. 34; 1.V.2, p. 101). In contemporary terminology, a proper name in the predicate position is, according to Mill, mentioned but not used. However, Lockwood tried to show that even within Mill's framework we can maintain that proper names in the predicate position do have connotation. Mill conceded that when we tell someone that a man is Brown we enable him to *identify* that individual (ibid., 1.II.5, p. 37). Thus, according to Lockwood (1975, pp. 494-5), what a proper name in the predicate position can be said to connote is *being identical with a certain individual*. I take Lockwood's analysis to be the same as mine, namely, that a proper name in the predicate position is used to say *who* someone is (or, in the case of places etc., *which* place some place is, etc.). Lockwood also argues effectively (§ II) against the very few arguments that can be found in Frege's writings and subsequent literature for the distinction between the 'is' of identity and that of predication.

The idea that identity statements are subject-predicate in form can easily be represented in the artificial language developed in this work. The basic form of sentences, according to this language, is '$(np_1, \ldots np_n)$ is/isn't P', where 'P' is an n-place predicate. In case $n=1$ and 'np_1' is a singular referring expression, the affirmative sentence is of the form 'a is P', where 'P' is a one-place predicate. We can now admit sentences of another form as well, those in which a singular referring expression 'b' replaces 'P' in the last formula: 'a is b'.

We now have to specify how these sentences are used in our deductive system. The following are their introduction and elimination rules:

---

natural language. What it does rather resembles the introduction of imaginary and complex numbers in order to prove some theorems about real numbers. By contrast, I use in my derivations only natural-language sentences, and all sentences in my arguments – not only their premises and conclusion – are supposed to be validly derived within natural language.

Other central aspects of Sommers' logic, as developed in his later book (1982), are also far removed from the analysis of natural language developed here. Sommers 'holds that reference begins with "some S"', and he 'treats most definite subjects (proper names, demonstratives, definite descriptions) as anaphoric expressions that have back reference to propositions of the form "some S is P"' (p. 5). Contrast section 6.2 above. He also thinks that every elementary sentence is of the form NP/VP, with *one* dominant noun phrase (e.g., p. 9); while elementary sentences, according to my analysis, may contain any number of noun phrases – none of which being more dominant, in some semantic sense, than any of the others – with an appropriate n-place predicate (cf. § 8.6).

Identity Introduction (Law of Identity). In any line (i) any sentence of the form '*a is a*' can be written, not relying on any line. Its justification is written 'IdI'.

Identity Elimination (Indiscerniblity of Identicals). Suppose that sentence (i) is '*a is b*', and that '*a*' appears in sentence (j) too. Then in any line (k) one can write the sentence identical to sentence (j) apart from the fact that in it *a* has been substituted by *b* in some or all of its appearances. Line (k) relies on the lines on which lines (i) and (j) rely. Its justification is written 'IdE, i, j'.

Given the meaning of identity, both derivation rules preserve consistency.

From IdI it follows that identity is reflexive. Let us now prove that it is also symmetric and transitive:

1	(1) *a is b*	Premise
	(2) *a is a*	IdI
1	(3) *b is a*	IdE, 2,1

1	(1) *a is b*	Premise
2	(2) *b is c*	Premise
1,2	(3) *a is c*	IdE, 2,1

Notice also that our derivation rules for identity cohere with our claim that the referential use of a proper name and of other referring expressions presupposes reference, not existence (p. 126). In any line in a proof we can write, for instance, 'Pegasus is Pegasus', where 'Pegasus' refers to the mythical horse which never really existed. And in contrast to the predicate calculus, we cannot infer from this by our derivation rules that Pegasus exists (i.e., we cannot infer the sentence which is usually translated by '(There is an x)(x = Pegasus)'). The possibility of such invalid arguments necessitates the addition of some constraints on the use of names to the deductive systems of the predicate calculus, constraints that might seem *ad hoc*. The system developed here needs no similar modifications.

Having allowed proper names into predicate position, we can allow anaphors into that position too, as in the sentence 'John is himself'. We can then prove the following:

1	(1) John is a man	Premise
	(2) John is John	IdI
	(3) John is himself	Anaphora Introduction, 2
	(4) Every man is himself	UI, 3,1

Since proper names and anaphors can occupy the predicate position, we might ask whether we should allow quantified noun phrases in this position as well. Indeed, sentences like 'John is every man' or 'Every man is some man' are exceptional. But notice, first, that the second sentence is both a tautology and a theorem of our system:

1	(1) John is a man	Premise
	(2) John is John	IdI
1	(3) John is some man	PI, 2,1
	(4) Every man is some man	UI, 3,1

Secondly, the use of quantified noun phrases in predicate position is natural and important in sentences that specify how many some particulars are. For instance, 'John and Peter are two men', 'They are five students', etc. It should therefore be permissible. In fact, we can even introduce a definition like the following:

> If 'a is a P', 'b is a P' and 'a isn't b' are true, then 'a and b are two P's' is also true, and *vice versa*.

This can be seen as a definition of being two. We can generalize it and give an inductive definition for all natural numbers:

> If 'a_1 and a_2 and ... and a_n are n P's', 'b is P', and 'b isn't a_1' and 'b isn't a_2' and ... 'b isn't a_n' are all true, then 'a_1 and a_2 and ... and a_n and b are $n+1$ P's' is also true, and *vice versa*.

We can now use these definitions to prove all sorts of arithmetical theorems, for instance, that two and two are four:

1 (1) a and b are two P's and c and d are two P's and a isn't c and a isn't d and b isn't c and b isn't d
Premise

1 (2) a and b are two P's and c is a P and d is a P and c isn't d and a isn't c and a isn't d and b isn't c and b isn't d sentence (1) and definition of 2

1 (3) a and b and c are three P's and d is a P and c isn't d and a isn't d and b isn't d
sentence (2) and definition of 3

1 (4) a and b and c and d are four P's   sentence (3) and definition of 4

Similar proofs can be given for the sum of any natural numbers. The commutativity and associativity of addition follow immediately from those of conjunction. In this way we may gain, by means of the deductive system developed above, a foothold in arithmetic. But although it contains hope, we better leave Pandora's Philosophy of Mathematics box shut, after having peeped inside through a crack.

Chapter 12

Conclusions

We have seen that the semantic categories of natural language do not coincide with those of the predicate calculus, and that some are implemented in them in different ways. Many, probably most sentences of natural language cannot be translated by sentences of the predicate calculus with the same, or even roughly the same semantic characteristics. These distinctions between natural language and the predicate calculus make the latter unfit for the analysis of the logic and semantics of the former.

This conclusion flies in the face of much of the tradition of analytic philosophy. Russell (1905), Reichenbach (1947, chap. VII), Quine (see the entry 'Predicate Logic' in his *Quiddities*), Davidson (1980, essays 6 and 7) and many others, including formal semanticists at least from Montague (1973; e.g., his translations on p. 266) onward, tried to analyze the meaning and logic of sentences of natural language, or sometimes of thoughts expressed by means of such sentences, with the help of some version of Frege's predicate calculus. One conclusion of this work is that this attempt is futile and should be abandoned.

Frege's own attitude to his calculus, inherited by some other philosophers, to a large extent originated this tradition.

Indeed, on the one hand Frege says that the calculus was developed to be a tool for carrying out chains of inferences, a task for which natural language is inadequate (1879, Preface, p. IV), and its application to arithmetic and geometry should therefore be most fruitful (ibid., p. VI; cf. 1884, § 91). His calculus was not intended for the analysis of natural language, but rather as a substitute for certain purposes (cf. Russell, 1957, pp. 387-8; Quine, 1953, pp. 150-1).

However, on the other hand, in order to justify his logic and calculus, Frege tries to show that contrary to what the grammar of natural language has led us to think, common nouns, *as used in natural language*, are not logical subjects but logical predicates. We saw in section 4.1 how he tried to establish the predicative nature of common nouns in such *natural language* sentences as 'All whales are mammals', 'All mammals have red blood', etc. He also considered his claim that concepts are functions one of his main achievements (Frege, 1906) – and among concept-words common nouns are meant to be included. Frege should therefore be taken to maintain that his calculus reveals in a perspicuous form the way concept-words function in the grammatically misleading natural language as well. For these reasons he should be considered the founder of the tradition that analyzes the semantics of natural language by means of the predicate calculus. In this book I tried to show that Frege was wrong in his analyses.

Russell's attitude resembles Frege's in this respect. In his 'On Denoting' of 1905 he analyzes sentences of English – e.g., 'I met a man', 'All men are mortal' and 'The father of Charles II was executed' – by translating them into a version of the predicate calculus. For that purpose he uses, for instance, variables, sentential functions and quantifiers as operators on sentential functions; and he interprets common nouns as predicates, and universal sentences as 'really hypothetical' (ibid., p. 481). Russell even criticizes Frege's treatment of denoting phrases by saying that it 'is plainly artificial, and does not give an exact analysis of the matter' (ibid., p. 484), demonstrating by that his commitment to a correct analysis of expressions of *natural language*. Later on, in his *Introduction to Mathematical Philosophy* of 1919, he again analyzes by means of the predicate calculus sentences of natural language, which include the four kinds of quantified sentences of Aristotelian logic and sentences with definite and indefinite descriptions (pp. 161-3 and chap. XVI). And he again claims that this analysis shows what their form 'really' is (p. 161) or what they 'contain' (p. 171). And he continued to hold until late in his life the same opinion on at least what his theory of descriptions achieves (1946, p. 859-60).

Like Frege, Russell evidently thinks that the predicate calculus reveals the way concepts function in natural language. According to him, the calculus departs from natural language in that syntactical ambiguity, which 'is hard to avoid in language', is 'easily avoided' in it (1905, p. 489). As late as 1957 (p. 388) he maintained that 'in philosophy, it is syntax, even more than vocabulary, that needs to be corrected.' He is therefore clearly in the tradition originated by Frege and criticized in this work, of analyzing the semantics and logic of natural language by means of the calculus. This tradition spread and established itself mainly through Russell's influential work.

In addition, Frege also writes that his new language will free thought from mistakes caused by natural language (1879, pp. VI-VII), that it is a language 'of pure thought' (ibid., subtitle), and that the new language is more adequate than natural language to express contents accurately (1882; 1883, p. 1). These last claims are also highly implausible in the light of this book: the predicate calculus is an impoverished language compared with natural language, and the latter seems capable of being as accurate as we need.

Lastly, Frege maintained (1879, p. VII) that logic has been too closely connected to language and grammar, and that one of the advantages of his calculus is that it will free logic from these connections. But in fact, what Frege brought about was that logic became closely connected to the grammar of a different language, his predicate calculus. *Pace* Frege, the logical relations that logic investigates are not between abstract Platonic thoughts but between sentences or utterances of a language. And a consequence of his *Begriffsschrift* was that instead of studying the logic of the languages we use, logicians studied the logic of a very different language, one which is hardly ever used anywhere outside logic.

In what may the value of an artificial language lie? It can be valuable if it enables us to study and analyze various aspects of natural language. The propositional calculus is such a language, representing some of the ways some sentence-connectives function in natural language. However, if the central claims made in this book are correct, the predicate calculus cannot serve this end.

On the other hand, an artificial language may *replace* natural language in certain domains, being a better tool to convey what we want to say about certain matters. The formulas of arithmetic constitute such a language, as well as musical notation. If we used only natural language in arithmetic or music, it would be practically impossible to convey what is easily expressed by means of those artificially devised languages. Moreover, one language can replace another even if they are very different, both semantically and logically. Accordingly, even if I am right and the study of natural language should not be conducted by means of the predicate calculus, one may still claim that the calculus should serve as the language of some science.

Frege indeed thought that his artificial language, the predicate calculus, would have such a function in mathematics (1879, p. VI; 1884, § 91). And some other philosophers have similarly maintained that the predicate calculus should replace natural language in the sciences, being a more adequate tool for science than the latter. Quine, for instance, claimed (1953, pp. 150-1) that formal logic, by which he meant the predicate calculus, is a tool for the scientist.

However, the development of mathematics has proved Frege mistaken: the predicate calculus is hardly ever used in mathematics outside mathematical logic. And since its invention, the calculus has not made any inroad into the sciences. Moreover, I am not acquainted with any good argument that explains why the predicate calculus should be more efficient than natural language for science. And lastly, given the impressive progress of the sciences in recent centuries, it is unclear why anyone should think that their languages need to be replaced. I therefore think there is no reason to hold the predicate calculus adequate as such a substitute, while there are several reasons for thinking it inadequate. In this respect the predicate calculus is of little or no value.

Some may argue that essential use is made of the predicate calculus in laying the foundations of arithmetic, namely in axiomatic set theory. I am not in a position to pass an adequate judgment on this claim. However, the predicate calculus was used in this project because it was the only powerful deductive system available. It is unknown whether any other system could also serve the same purpose. And my definitions of quantities and my proof that two and two are four in section 11.5 (p. 147) suggest perhaps that my system may also be used for this purpose. Secondly, a theory which identifies the number one, say, with the set whose only member is the set which has no member, can hardly be considered as the real foundation of the claim that one and one are two. One is rather reminded of Wittgenstein's remark, that 'the *mathematical* problems of what is called foundations are no more the foundation of mathematics for us than the painted rock is the support of a painted tower' (1978, p. 378). I am therefore doubtful about whether its use in axiomatic set theory can prove the predicate calculus essential for laying the foundations of arithmetic.

However, in case the calculus is found essential, or most convenient, to this purpose, this will endow it with a very important function, although one which is considerably limited relative to those currently ascribed to it.

It may still be said in support of the logical value of the predicate calculus that it is a very good tool for carrying out inferences. Moreover, because of this, it made possible the study of properties of inferences and of deductive systems: consistency, completeness, decidability, compactness, and so on. The work of Frege, Russell, Gödel, Gentzen, Tarski and others has been invaluable in this respect, and they all relied on some version of the predicate calculus. We should not commit it then to the flames.

I agree that the predicate calculus is suitable for carrying out complex inferences, and that for this reason it has been used in the important logical inquiries just mentioned. But in the third part of this work I developed a deductive system for natural language, comparable in its deductive power to the first order predicate calculus. Logicians' interest in a system of formal inferences with propositional combinations of multiply quantified sentences may be served by that system as well.

Moreover, the just mentioned logical properties of inferences in a language – consistency, completeness, etc. – are defined independently of any language, and can be illustrated in the propositional calculus. The predicate calculus is not essential for that purpose. The fact that historically, the predicate calculus supplied the initiative for many logical inquiries and served to illustrate them, does not entail that it is essential for that purpose and should therefore be preserved.

In addition, if the expressive power of the calculus is limited, then its advantages as a formal system are insufficient to recommend it as a substitute for natural language for carrying out inferences. After all, the fact that the propositional calculus has, as a system of formal logical inferences, the advantage over the predicate calculus of being decidable, is insufficient to make it preferable over the latter for that task; and that is precisely because its expressive power is very limited compared with that of the predicate calculus. We should use, for our inferences, a language in which we can say what we need to say; and while natural language is such a language, the predicate calculus is not.

Lastly, a system's superiority as a formal system of inferences is insufficient to recommend it as a tool for actually carrying out inferences. (And this applies to the predicate calculus as well as to my formal system.) When we actually make inferences in the course of our scientific and other inquiries, they are rarely formal. Moreover, when a real doubt concerning the validity of a certain form of inference arises, we hardly ever check its validity by consulting any formal system. Rather, we use examples and models to establish the validity or invalidity of the argument form. Now our ability to carry out inferences in natural language, in view of the enormous progress of our sciences, is indisputable. It is therefore doubtful whether any advantage as a *formal* system of inferences is significant when our ability to carry out inferences is considered.

If the predicate calculus has no significant use, then an investigation of its distinctive properties is not of much interest. The calculus may still maintain some formal interest, since it is distinct from the propositional calculus, and from other calculi, in some logical properties: while the propositional calculus is complete and decidable, the predicate calculus is complete but not decidable. In this respect it is also unlike the second-order predicate calculus, which is neither complete nor decidable. (I have not yet inquired whether the deductive system developed in this book is either complete or decidable.) It can thus serve to illustrate this combination of logical properties. This value, however, is very limited, and can hardly justify the effort required to master the artificial language of the predicate calculus.

Bibliography

ברגמן, שמואל הוגו (1975), *מבוא לתורת ההגיון*, מהדורה שלישית, מוסד ביאליק, ירושלים.

ברי, נמרוד (1978, תשל"ח), 'שם-תואר מועצם ושם-עצם מותאר בעברית חדשה דיבורית', *לשוננו*, **42**, עמודים 252-72.

Al-Fārābī (1956), *Eisagoge*, edited and translated by D.M. Dunlop, *The Islamic Quarterly*, **3**, pp. 117-38.

Altham, J.E.J. and Tennant, N.W. (1975), 'Sortal Quantification', in Edward L. Keenan (ed.), *Formal Semantics of Natural Language*, Cambridge University Press, pp. 46-58.

Aquinas, St. Thomas (1965), *Summa Theologiae*, Volume VI, Latin Text and English Translation, Blackfriars.

Aristotle (1941), *Categories, On Interpretation, Prior Analytics*, in R. McKeon (ed.), *The Basic Works of Aristotle*, Random House, New York.

Armstrong, D.M. (1978), *Nominalism and Realism: Universals and Scientific Realism, Volume I*, Cambridge University Press.

Barker, S.J. (1997), 'E-Type Pronouns, DRT, Dynamic Semantics and the Quantifier/Variable-Binding Model', *Linguistics and Philosophy*, **20**, pp. 195-228.

Barri, N. (1978), See Hebrew entry at top.

Barwise, J. (1979), 'On Branching Quantifiers in English', *Journal of Philosophical Logic*, **8**, pp. 47-80.

Barwise, J. and Cooper, R. (1981), 'Generalized Quantifiers and Natural Language', *Linguistics and Philosophy*, **4**, pp. 159-219.

Ben-Yami, H. (2001), 'The Semantics of Kind Terms', *Philosophical Studies*, **102**, pp. 155-84.

Bergmann, Shmuel Hugo (1975), See Hebrew entry at top.

Black, Max (1971), 'The Elusiveness of Sets', *The Review of Metaphysics*, **24**, pp. 614-36.

Boolos, G. (1984), 'To Be is to Be a Value of a Variable (or to Be Some Values of Some Variables)', reprinted in his (1998), *Logic, Logic, and Logic*, Harvard University Press, pp. 54-72.

Boolos, G. (1985), 'Nominalist Platonism', reprinted in his (1998), *Logic, Logic, and Logic*, Harvard University Press, pp. 73-87.

Bradley, F.H. (1922), *The Principles of Logic*, corrected impression of the 2nd edition, Oxford University Press.

Buridan, Jean (1966), *Sophisms on Meaning and Truth (Sophismata)*, translated by Theodore Kermit Scott, Appleton-Century-Crofts, New York.

Cameron, J.R. (1999), 'Plural Reference', *Ratio (new series)* **12**, pp. 128-47.

Castañeda, H.N. (1967), 'Comments on D. Davidson's "The Logical Form of Action Sentences"', in N. Rescher (ed.), *The Logic of Decision and Action*, University of Pittsburgh Press, pp. 104-12.

Davidson, Donald (1967), 'The Logical Form of Action Sentences', reprinted in his (1980), pp. 105-48.

Davidson, Donald (1980), *Essays on Actions and Events*, Clarendon, Oxford.

De Morgan, Augustus (1847), *Formal Logic*, Taylor and Walton, London.

Donnellan, K. (1966), 'Reference and Definite Descriptions', *The Philosophical Review*, **75**, pp. 281-304.

Dummett, Michael (1981), *Frege: Philosophy of Language*, 2nd edition, Duckworth, London.

Dummett, Michael (1991), *Frege: Philosophy of Mathematics*, Duckworth, London.

Evans, Gareth (1977a), 'Pronouns, Quantifiers, and Relative Clauses (I)', reprinted in his (1985), pp. 76-152.

Evans, Gareth (1977b), 'Pronouns, Quantifiers, and Relative Clauses (II)', reprinted in his (1985), pp. 153-75.

Evans, Gareth (1980), 'Pronouns', reprinted in his (1985), pp. 214-48.

Evans, Gareth (1982), *The Varieties of Reference*, Oxford University Press.

Evans, Gareth (1985), *Collected Papers*, Oxford University Press.

Frege, Gottlob (1879), *Begriffsschrift: Eine der Arithmetischen nachgebildete Formelsprache des reinen Denkens*, Verlag von Louis Nebert, Halle A/S.

Frege, Gottlob (1882), 'Ueber die wissenschaftliche Berechtigung einer Begriffsschrift', *Zeitschrift für Philosophie und philosophische Kritik*, **81**, pp. 48-56.

Frege, Gottlob (1883), 'Ueber den Zweck der Begriffsschrift', *Sitzungsberichte der Jenaischen Gesellschaft für Medizin und Naturwissenschaft für das Jahr 1882*, Verlag von G. Fischer, Jena, pp. 1-10.

Frege, Gottlob (1884), *Die Grundlagen der Arithmetik*, Wilhelm Koebner, Breslau.

Frege, Gottlob (1891), 'Funktion und Begriff', Hermann Pohle, Jena.

Frege, Gottlob (1892), 'Über Begriff und Gegenstand', *Vierteljahrsschrift für wissenschaftliche Philosophie*, **16**, pp. 192-205.

Frege, Gottlob (1893 and 1903), *Grundgesetze der Arithmetik*, selections translated in P. Geach and M. Black (1960), *Translations from the Philosophical Writings of Gottlob Frege*, Blackwell, Oxford.

Frege, Gottlob (1894), 'Review of Husserl's *Philosophie der Arithmetik*', *Zeitschrift für Philosophie und phil. Kritik*, **103**, pp. 313-32, selections translated in P. Geach and M. Black (1960), *Translations from the Philosophical Writings of Gottlob Frege*, Blackwell, Oxford.

Frege, Gottlob (1895), 'Kritische Beleuchtung einiger Punkte in E. Schröders *Vorlesungen über die Algebra der Logik*', *Archiv für systematische Philosophie*, **1**, pp. 433-56.

Frege, Gottlob (1897), 'Logic', in his (1979), *Posthumous Writings*, translated by Peter Long and Roger White, Blackwell, Oxford, pp. 126-51.

Frege, Gottlob (1906), 'Was kann ich als Ergebnis meiner Arbeit ansehen?', in his (1969), *Nachgelassene Schriften*, Felix Meiner, Hamburg, p. 200.

Frege, Gottlob (1914), 'Logic in Mathematics', in his (1979), *Posthumous Writings*, translated by Peter Long and Roger White, Blackwell, Oxford, pp. 203-50.

Frege, Gottlob (1976), *Wissenschaftlicher Briefwechsel*, Felix Meiner, Hamburg.

Geach, P.T. (1956), 'Good and Evil', *Analysis*, **17**, pp. 33-42.

Geach, P.T. (1961-62), 'Namely-Riders Again', reprinted in his (1972), *Logic Matters*, Blackwell, Oxford, pp. 92-5.

Geach, P.T. (1962), *Reference and Generality: An Examination of Some Medieval and Modern Theories*, emended edition 1968, Cornell University Press.

Geach, P.T. (1968), 'History of the Corruptions of Logic', reprinted in his (1972), *Logic Matters*, Blackwell, Oxford, pp. 44-61.

Gentzen, Gerhard (1934-5), 'Untersuchungen über das logische Schliessen', *Mathematische Zeitschrift*, **39**, pp. 176-210, 405-31.

Grice, H.P. (1967), 'Logic and Conversation', reprinted in his (1989), *Studies in the Way of Words*, Harvard University Press, pp. 1-143.

Grice, H.P. and Strawson, P.F. (1956), 'In Defense of a Dogma', *The Philosophical Review*, **65**, pp. 141-58.

Hale, B. (1987), *Abstract Objects*, Blackwell, Oxford.

Higginbotham, J. (1998), 'On Higher-Order Logic and Natural Language', *Proceedings of the British Academy*, **95**, pp. 1-27.

Hülsen, Reinhard (2000), 'Understanding the Semantics of "relativa grammaticalia": Some Medieval Logicians on Anaphoric Pronouns', in K. von Heusinger and U. Egli (eds), *Reference and Anaphoric Relations*, Kluwer, pp. 31-46.

van Inwagen, Peter (1990), *Material Beings*, Cornell University Press.

Keenan, E.L. (1992), 'Beyond the Frege Boundary', *Linguistics and Philosophy*, **15**, pp. 199-221.

Keenan, E.L. (1996), 'The Semantics of Determiners', in S. Lappin (ed.), *The Handbook of Contemporary Semantic Theory*, Blackwell, Oxford, pp. 41-63.

Keenan, E.L. and Westerståhl, Dag (1997), 'Generalized Quantifiers in Linguistics and Logic', in Johan van Benthem and Alice ter Meulen (eds), *Handbook of Logic and Language*, Elsevier and The MIT Press, pp. 837-93.

Kenazawa, Makoto (2001), 'Singular Donkey Pronouns Are Semantically Singular', *Linguistics and Philosophy*, **24**, pp. 383-403.

Kolaitis, P. and Väänänen, J. (1995), 'Generalized Quantifiers and Pebble Games on Finite Structures', *Annals of Pure and Applied Logic*, **74**, pp. 23-75.

Kripke, Saul (1980), *Naming and Necessity*, Blackwell, Oxford.

Lemmon, E.J. (1965), *Beginning Logic*, Nelson, London.

Leonard, H. and Goodman, N. (1940), 'The Calculus of Individuals and Its Uses', *Journal of Symbolic Logic*, **5**, pp. 45-55.

Lewis, David (1991), *Parts of Classes*, Blackwell, Oxford.

Lockwood, Michael (1975), 'On Predicating Proper Names', *The Philosophical Review*, **84**, pp. 471-98.

Lønning, Jan Tore (1997), 'Plurals and Collectivity', in Johan van Benthem and Alice ter Meulen (eds), *Handbook of Logic and Language*, Elsevier and The MIT Press, pp. 1009-53.

McCawley, J.D. (1968), 'The Role of Semantics in Grammar', in E. Bach and R.T. Harms (eds), *Universals in Linguistic Theory*, Holt, Rinehart and Winston, New York, pp. 124-69.

Mill, J.S. (1872), *A System of Logic*, eighth edition, London.

Montague, R. (1973), 'The Proper Treatment of Quantification in ordinary English', reprinted in his (1974), *Formal Philosophy*, Yale University Press, pp. 247-70.

Morton, Adam (1975), 'Complex Individuals and Multigrade Relations', *Noûs*, **9**, pp. 309-18.

Neale, Stephen (1990), *Descriptions*, The MIT Press.

Neale, Stephen (1995), 'The Philosophical Significance of Gödel's Slingshot', *Mind*, **104**, pp. 761-825.

Newton-Smith, W.H. (1985), *Logic: An Introductory Course*, Routledge, London.

Oliver, Alex and Smiley, Timothy (2001), 'Strategies for a Logic of Plurals', *The Philosophical Quarterly*, **51**, pp. 289-306.

Peacocke, C. (1979), 'Game-Theoretic Semantics, Quantifiers and Truth: Comments on Professor Hintika's Paper', in E. Saarinen (ed.), *Game-Theoretical Semantics*, Dordrecht, Holland, pp. 119-34.

Platts, M. de Bretton (1979), *Ways of Meaning*, Routledge and Kegan Paul, London.

Quine, W.V. (1948), 'On What There Is', reprinted in his (1961), *From a Logical Point of View*, second edition, Harvard University Press, pp. 1-19.

Quine, W.V. (1953), 'Mr. Strawson on Logical Theory', reprinted in his (1976), *The Ways of Paradox and Other Essays*, second edition, Harvard University Press, pp. 139-57.

Quine, W.V. (1960), *Word and Object*, The MIT Press.

Quine, W.V. (1981), 'Predicates, Terms and Classes', in his *Theories and Things*, Harvard University Press, pp. 164-72.

Quine, W.V. (1982), *Methods of Logic*, fourth edition, Harvard University Press.

Quine, W.V. (1987), *Quiddities: An Intermittently Philosophical Dictionary*, Penguin Books.

Quine, W.V. (1995), *From Stimulus to Science*, Harvard University Press.

Reichenbach, Hans (1947), *Elements of Symbolic Logic*, The Macmillian Company, New York.

Rescher, N. (1962), 'Plurality-quantification', abstract, *Journal of Symbolic Logic*, **27**, pp. 373-4.

Russell, Bertrand (1903), *The Principles of Mathematics*, second edition 1937, George Allen and Unwin, London.

Russell, Bertrand (1905), 'On Denoting', *Mind*, **56**, pp. 479-93.

Russell, Bertrand (1919), *Introduction to Mathematical Philosophy*, George Allen and Unwin, London.

Russell, Bertrand (1946), *History of Western Philosophy*, George Allen and Unwin, London.

Russell, Bertrand (1948), *Human Knowledge, Its Scope and Limits*, George Allen and Unwin, London.

Russell, Bertrand (1957), 'Mr. Strawson on Referring', *Mind*, **66**, pp. 385-9.

Schein, B. (1993), *Plurals and Events*, The MIT Press.

Simons, Peter (1982), 'Number and Manifolds' and 'Plural Reference and Set Theory', in Barry Smith (ed.), *Parts and Moments: Studies in Logic and Formal Ontology*, Philosophia Verlag, München and Wien, pp. 160-260.

Sommers, Fred (1969), 'Do We Need Identity?', *The Journal of Philosophy*, **66**, pp. 499-504.

Sommers, Fred (1982), *The Logic of Natural Language*, Clarendon, Oxford.

Strawson, P.F. (1950), 'On Referring', reprinted in his (1971), pp. 1-27.

Strawson, P.F. (1952), *Introduction to Logical Theory*, Methuen, London.

Strawson, P.F. (1964), 'Identifying Reference and Truth-Values', reprinted in his (1971), pp. 75-95.

Strawson, P.F. (1967), 'Is Existence Never a Predicate?', reprinted in his (1974), *Freedom and Resentment and Other Essays*, Methuen, London, pp. 189-97.

Strawson, P.F. (1970), 'The Asymmetry of Subjects and Predicates', reprinted in his (1971), pp. 96-115.

Strawson, P.F. (1971), *Logico-Linguistic Papers*, Methuen, London.

Strawson, P.F. (1974), *Subject and Predicate in Logic and Grammar*, Methuen, London.

Strawson, P.F. (1986), 'Direct Singular Reference: Intended Reference and Actual Reference', reprinted in his (1997), *Entity and Identity*, Clarendon, Oxford, pp. 92-9.

Weidemann, H. (1986), 'The Logic of Being in Thomas Aquinas', in S. Knuuttila and Jaakko Hintikka (eds), *The Logic of Being*, Reidel, Dordrecht, pp. 181-200.

Westerståhl, Dag (2001), 'Quantifiers', in Lou Goble (ed.), *The Blackwell Guide to Philosophical Logic*, Blackwell, Oxford, pp. 437-60.

Wiggins, D. (1981), '"Most" and "All": Some Comments on a Familiar Programme, and on the Logical Form of Quantified Sentences', in Mark Platts (ed.), *Reference, Truth and Reality*, Routledge and Kegan Paul, London, pp. 318-46.

Wiggins, D. (1997), 'Meaning and Truth Conditions: from Frege's Grand Design to Davidson's', in B. Hale and C. Wright (eds), *A Companion to the Philosophy of Language*, Blackwell, Oxford, pp. 3-28.

Wittgenstein, Ludwig (1953), *Philosophical Investigations*, translated by G.E.M. Anscombe, second edition 1958, Blackwell, Oxford.

Wittgenstein, Ludwig (1978), *Remarks on the Foundations of Mathematics*, third edition, Blackwell, Oxford.

Yi, Byeong-uk (1999), 'Is Two a Property?', *The Journal of Philosophy*, **96**, pp. 163-90.

Index

adjectives 8, 36-7, 41, 44-5, 62, 65, 79, 116
    attributive 32-5, 62, 136
    predicative 32-4
Al-Fārābī 144
anaphora 3, 30, 96-100, 118, 120-23,
    145-6
    bound 3, 64, 82, 86, 96-8, 100-110,
        120-23, 125, 136-8, 141
    donkey 96-7, 100-107, 110, 120,
        138-40
Aquinas, T. 63-4, 144
arithmetic 42, 147-8, 150; *see also*
    mathematics; number
Aristotle 37, 41, 63, 65, 92, 129-31, 143;
    see also Aristotelian Logic
Aristotelian logic 3, 8, 28, 39, 41, 62, 67,
    73, 78, 110-11, 128-33, 139, 149;
    see also Aristotle
artificial language 44, 61, 80, 82-3, 91,
    116, 145, 150-52

Barwise, J. 68, 71-2, 88
Black, M. 10-13, 50
Boolos, G. 10, 26, 85-7
Bradley, F. H. 44-6
Buridan, J. 101-2
    Buridan's Law 52, 62

Castañeda, H. N. 24 6
common nouns
    claimed to be always predicative ix,
        8, 12, 41-4, 55, 87, 148-9
    classificatory use of 36-7, 65
    empty 37-9
    as predicates 36-7, 39, 64-5
    as referring expressions 2, 8-9,
        11-13, 28-32, 34-5, 37, 39, 43,
        54-5, 62, 64-5, 67, 126
converse relation-names 2, 91
Copper, R., *see* Barwise, J.
copula 2, 62, 64, 80, 83, 91-4, 110-11,
    116, 118, 137

alleged ambiguities 31-2, 143-4

Davidson, D. 23-6, 61, 77, 148
De Morgan, A. 44, 67, 139
deductive systems 151
    mechanical versus formal 117
    for natural language 54, 109, 115,
        137, 144
    the one developed in this book 2-3,
        64, 73, 75-6, 85, 101, 107,
        109-10, 115-42, 145-7, 150-52
    of the predicate calculus 115-6, 144,
        146, 148-51
definite descriptions 9-13, 16, 19-20, 25,
    28-30, 51-3, 62, 95, 103, 142, 145,
    149
demonstratives 7, 9, 11, 16, 18, 25-6,
    28-30, 87, 145
domain of discourse 1, 3, 28, 39, 59-61,
    66, 70-71, 126
Dummett, M. 28, 55, 78

Evans, G. 7, 68, 96-7, 100
existence 19, 37-8, 73-5
    'exist' a predicate 40, 73
    presupposition of 39-40, 126, 146;
        see also reference,
        presupposition of

fictional characters 39-40, 53-4, 126; *see
    also* mythological characters;
    reference, presupposition of
Frege, G. 1-3, 8, 12, 18, 20-25, 28-9,
    31-2, 34, 37, 39, 41-4, 55, 60, 64,
    67-8, 73, 75, 87, 90, 92, 142-5,
    148-51
    Begriffsschrift 1, 8, 28, 31, 42, 44,
        73, 143, 148-50
    Grundlagen 22, 28, 42-3, 68, 148, 150
    'Kritische Beleuchtung' 32, 43-4, 154
    'Über Begriff und Gegenstand' 8, 32,
        42-3, 143-4

Geach, P. 10, 33, 52, 62-6, 107
Geach-Kaplan sentence 107
Gentzen, G. 117, 151
Grice, P. 49, 75, 99

Higginbotham, J. 25-7, 78

identifying 18, 35-6, 51-2, 142, 145; *see also* identity
identity 3, 51-2, 133, 142-6; *see also* identifying
immediate inferences 3, 128-9, 130-31
indefinite article 103-7, 149
van Inwagen, P. 12-13

Kripke, S. 35-6

Lockwood, M. 145

mathematics 1, 41-2, 44, 67-8, 144, 150; *see also* arithmetic; number
McCawley, J. D. 24
Mill, J. S. 35, 41, 145
model-theoretic semantics 3, 23, 31
Montague, R. 67, 148
Morton, A. 12-13
multiply quantified sentences, *see* quantification, multiple
mythological characters 40, 146; *see also* fictional characters; reference, presupposition of

natural deduction, *see* deductive systems
natural kind terms 35-7
Neale, S. 7, 19-20, 102-3, 106
noun phrases 9, 14, 26, 33, 69, 72, 81-2, 85, 144-5
    anaphoric 96-100, 104, 106, 109-10, 118, 138; *see also* anaphora
    definite 30, 61-3, 81-2, 85, 87-9, 92, 95-8, 107, 109, 115, 117-24, 133-5
    E-type 100-104, 108, 137
    indefinite 61, 63, 104
    quantified 11, 13, 28-46 *passim*, 55, 59-60, 62, 78-127 *passim*, 133-41 *passim*, 146-7
number 13-15, 147, 150

Oliver, A. 23-7

passive voice 2, 88, 90, 116
predicate calculus 1-4, 7-8, 10, 12-13, 16-21, 23, 26, 28, 32-4, 37-8, 41-4, 59-61, 64, 66-8, 71-6, 78-81, 86, 89-98, 100, 104-6, 108, 110, 115-8, 120, 126, 129, 132, 135-6, 139, 146, 148-52
predicates 3, 7-8, 12-14, 18-19, 26, 31-7, 39-44, 54, 60, 62-5, 70-72, 76, 78, 80, 82-3, 92, 98, 109-10, 115, 117-8, 128, 130-31, 133-6, 142-7, 149; *see also* predication
    conjunction and disjunction of 15-16, 107-9, 136-8
predicate-variables 79, 82-3, 117
predication 2-3, 8-9, 12-13, 31-2, 36, 39, 41, 44, 53, 60, 62, 64-5, 92-3, 110-11, 115-6, 142-5; *see also* predicates
    collective 21-6, 84-5, 88-9, 106-7, 115
    distributive 21-3, 25, 27, 84-5, 88-9, 106-7, 110, 115
    minimalist conception of 64-5, 142-3, 145
pronouns 7, 9, 18, 51, 97, 99-100, 105
    E-type 100-103, 108, 137
    plural 9, 11, 16, 25, 28-30, 87
    as variables 64, 95-8; *see also* anaphora, bound
proper names 3, 7-10, 12, 18, 21-2, 26, 35, 37, 43, 49, 72, 87, 95-6, 110, 120, 143-6
    in predicate position 36-7, 143, 145-6
propositional calculus 61, 117-8, 129, 131, 136, 150-52

quantification 2-3, 10, 12-3, 19, 21, 25, 28, 31, 59-141 *passim*, 149; *see also* noun phrases, quantified; quantifiers
    cumulative 89
    iterative 80-84, 87-91, 109-10
    multiple 2-3, 64, 78-94, 133-6, 138-41, 151

parallel 88-9
quantifiers 1, 14, 29-30, 39, 46, 59-72,
    76, 80, 82-3, 85, 91-4, 97,
    103-111, 115, 135-6, 149; *see also*
    noun phrases, quantified;
    quantification; indefinite article
  binary 19, 39, 60, 66-72, 97
  comparative 69-72
  complementary 94
  conservative 70
  existential 73-5
  generalized 2, 39, 68, 91-2
  particular 73-5, 107, 122-5, 133,
    135, 138
  restricted 60, 66-72, 91, 97-8, 104
  universal 63, 104-7, 111, 120-22,
    124-7, 133-5, 141
Quine, W. V. 7-8, 47-9, 61, 95-6, 107,
    148, 150

reference
  to future individuals 31, 45-6, 53
  the nature of 2, 7-12, 29, 45-55
  to past individuals 31, 53
  plural, the nature of 2, 9-16, 47,
    50-52, 54-5
  and pointing 11, 18, 40, 46, 48-53,
    62
  presupposition of 38-40, 74-5, 88,
    124-6, 139, 146; *see also*
    existence, presupposition of;
    fictional characters;
    mythological characters
relations 3, 12-13, 39, 60-61, 75-7,
    90-91, 119, 133, 140-41
Russell, B. 3, 19-22, 24, 29, 35, 44-5,
    87, 144, 148-9, 151

Schein, B. 25-7
science 150-51
semantically derived nouns 54-5, 111-2
semantically isomorphic translation
    17-20, 28, 80
sentential functions 1, 60-61, 66, 68,
    97-8, 108, 110, 136, 149
Simons, P. 11-14, 29
Smiley, T., *see* Oliver, A.
Sommers, F. 144-5
Square of Opposition 3, 128-31
Strawson, P. 10, 15-16, 19-20, 39-40,
    49, 62-3, 65, 73, 95, 109, 128, 131
subject-predicate sentences 31-2, 38, 40,
    90, 142-5
subjects, grammatical or logical 8,
    13-15, 19, 21-2, 24, 26, 28, 32-4,
    37-9, 41-4, 54, 62-4, 76, 78, 80,
    85, 87, 92-3, 101, 110-11, 115-6,
    118, 128, 130-36, 143-5, 148
  disjunctive 9, 12, 15-16, 21, 28
syllogisms 3, 41, 67, 115, 128-9, 131,
    139

Tarski, A. 3, 151; *see also* Tarskian
Tarskian 87, 110; *see also* Tarski, A.
Transposition 119, 133-6

variables 79-80, 82-3, 117
  plural 12-13
  in the predicate calculus 1, 3, 42, 45,
    60-61, 64, 71, 87, 95-100, 110, 149

Wiggins, D. 1, 68
Wittgenstein, L. 17, 48-50, 55, 150

Yi, B. 13-1

For Product Safety Concerns and Information please contact our EU
representative GPSR@taylorandfrancis.com
Taylor & Francis Verlag GmbH, Kaufingerstraße 24, 80331 München, Germany